SORGHUM
An Emerging Cash Crop

B. Dayakar Rao
J.V. Patil
K. Nirmal Reddy
Vinay K. Soni
G. Srivatsava

FOUNDATION®
BOOKS

FOUNDATION®
BOOKS

Foundation Books is an imprint of Cambridge University Press India Pvt. Ltd.

Cambridge House, 4381/4 Ansari Road, Daryaganj, New Delhi 110 002

www.cambridgeindia.org

First published 2014

Printed in India at Replika Press Pvt. Ltd.

ISBN 978-93-82993-52-0 Paperback

Contents

List of Figures

Annexure figures

List of Tables

Annexure Tables

Foreword I

Sorghum is a traditional staple food in India. The consumption of sorghum both in urban and in rural areas has been on the decline for various reasons – easy access of other cereals like rice and wheat, tedious and time-consuming processes involved in preparation, urbanisation, and Government policies to supply fine cereals at a subsidized rate. However, sorghum is preferred by nutritionists over other cereals for its high fibre content and slow-releasing carbohydrates and minerals. Thus, it facilitates in dietary management and in preventing metabolic diseases. It is highly recommended for all age groups especially for those suffering from lifestyle diseases such as diabetics and obesity. Therefore, there is a need for diversification of processing technology that makes sorghum products both value added and convenient.

The current study is part of a series of outputs of the National Agricultural Innovation Project (NAIP) focusing on Millets Value Chain headed by the Directorate of Sorghum Research (DSR). Sorghum and millets are the future crops offering nutritional security owing to their nutritional superiority over other fine cereals such as rice and wheat. The consumer acceptance report of Hyderabad, Pune, and Delhi shows that sorghum products are quite popular in market. This book discusses the status of sorghum production and its consumption in India. It also talks about the strategies to exploit the nutritional value of sorghum, and market point of view. I thank the authors for their sincere effort to write this book.

S. Ayyappan
Director General
Indian Council for Agricultural Research (ICAR)

Foreword II

Sorghum is a drought-tolerant crop and thus an important food source in semi-arid regions of the world. The area under sorghum production has declined from 18 million ha in 1970 to 6.98 million ha in 2009–10. This decline is more prominent in *kharif* (rainy) season than in *rabi* (post-rainy) season. The consumer demands in food production have changed considerably in the last three decades. It is believed that food is not only required to provide necessary nutrients but also to prevent nutrition-related diseases and improve physical and mental well-being of consumers. Sorghum products are being promoted for their positive health benefits by the Directorate of Sorghum Research under the National Agricultural Innovation Project (NAIP). This promotion aims at rural income augmentation and employment generation through post-harvest processing, value addition, and development of rural agro-industries.

I appreciate the collaborative efforts of the authors in bringing out this valuable publication. It will be an important reference book for planners, policy-makers, plant breeders, socio-economists, and nutritionists alike to devise steps to enhance the production, consumption, utilization, and marketing of sorghum.

Swapan Kumar Datta
Deputy Director General (Crop Sciences)
Indian Council for Agricultural Research (ICAR)

Preface

Traditionally, ready-to-eat (RTE) and ready-to-cook (RTC) products are made from fine cereals such as rice and wheat. There has been no systematic effort to make these products from coarse cereals such as *jowar*, *bajra*, and *ragi* due to a variety of issues related to RTE and RTC products, their processing, and the technology involved. Besides, the health benefits of the coarse cereals have not been communicated to the new age consumers effectively.

The Directorate of Sorghum Research (DSR), Rajendranagar, Hyderabad, conducted a project on 'Creation of demand for millet foods through production to consumption system value chain' with the financial assistance of Indian Council of Agricultural Research (ICAR)/National Agricultural Innovation Project (NAIP). Consequently, the DSR has conducted a lot of research over the years and has successfully overcome product-/process-related barriers/issues through innovation. It has brought out over 20-odd products made from sorghum. Out of these products, five most popular products (*jowar*-rich multigrain *atta*, *jowar* biscuits, *jowar* vermicelli, *jowar suji*/*rawa*/semolina, and *jowar* flakes) have been chosen for launch in the market. The processing technologies for these five products have been perfected to commercial scale. Before inviting private entrepreneurs, DSR initiated a demand assessment and consumer acceptability study for these products. The impact of the major outcome of the study on consumer acceptability of these five products has been documented along with an estimation of demand and the potential marketing and pricing strategy.

The consumer survey has been conducted in a professional manner and the proposed strategies based on the results are the outcome of rigorous brainstorming. We hope this book will be useful for all existing and upcoming entrepreneurs who are looking at manufacturing and marketing of *jowar*-based food.

Acknowledgements

The authors wish to acknowledge the National Agricultural Innovation Project (NAIP) and the authorities for providing financial assistance to conduct this research and document it in the form of this book. The institutional support rendered by Directorate of Sorghum Research (DSR) is also acknowledged with gratitude. The help received from various sources in soliciting the relevant information from public and private agencies is appreciated. The authors are also grateful to M/s Commodity India Ltd for their efforts in giving final shape to this book. They are indebted to Professor H.S. Talwar, Principal Scientist, DSR, Hyderabad, for taking out time to review the manuscript. The authors also appreciate the help rendered by the CAC members, Dr Bangali Baboo, former National Director and late Dr R.K. Goyal, National Coordinator, NAIP; Dr D. Rama Rao, ND, and Dr Ezekiel, NC II, PIU; and the project staff at DSR.

List of Abbreviations

CHPT_A	Appearance of *jowar chapati/roti*
CHPT_CLR	Colour of *jowar chapati/roti*
CHPT_OAL	Overall rating of *jowar chapati/roti*
CHPT_SFT	Softness of *jowar chapati/roti*
CHPT_SML	Smell of *jowar chapati/roti*
CHPT_SWL	Ability of *jowar chapati/roti* to swell
CHPT_TST	Taste of *jowar chapati/roti*
CHPT_TX	Texture of *jowar chapati/roti*
JBS_CLR	Colour of *jowar* biscuit
JBS_CRP	Crispiness of *jowar* biscuit
JBS_OAL	Overall rating of *jowar* biscuit
JBS_SML	Smell of *jowar* biscuit
JBS_TST	Taste of *jowar* biscuit
JFL_APR	Raw appearance of *jowar* flakes/*poha*
JFL_PRP	Preparation process of *jowar* flakes/*poha*
JFL_TST	Taste of *jowar* flakes/*poha*
JMAT_APR	Raw appearance of *jowar*-rich multigrain *atta*
JMAT_PRP	Preparation process of *jowar*-rich multigrain *atta*
JMAT_TST	Taste of *jowar*-rich multigrain *atta*
JOAT_CLR	Colour attribute of *jowar*-rich multigrain *atta*
JOAT_ERL	Ease of rolling out *roti* attribute of *jowar*-rich multigrain *atta*
JOAT_OAL	Overall attribute of *jowar*-rich multigrain *atta*
JOAT_SFD	Softness of dough attribute of *jowar*-rich multigrain *atta*
JOAT_SML	Smell attribute of *jowar*-rich multigrain *atta*
JOAT_TX	Texture attribute of *jowar*-rich multigrain *atta*
JOBS_A	Appearance of *jowar* biscuit
JOBS_CLR	Colour of *jowar* biscuit
JOBS_CRS	Crispiness of *jowar* biscuit
JOBS_FLT	Fullness of test after eating *jowar* biscuit
JOBS_MFL	Mouth feel of *jowar* biscuit
JOBS_OAL	Overall rating of *jowar* biscuit

JOBS_SML	Smell of *jowar* biscuit
JOBS_SWT	Sweetness/saltiness of *jowar* biscuit
JOBS_THK	Thickness of *jowar* biscuit
JOFL_A	Appearance of *jowar* flakes/*poha*
JOFL_CLR	Colour of *jowar* flakes/*poha*
JOFL_OAL	Overall rating of *jowar* flakes/*poha*
JOFL_SML	Smell of *jowar* flakes/*poha*
JOFL_THK	Thickness of *jowar* flakes/*poha*
JOSJ_A	Appearance of *jowar suji*/*rawa*/semolina
JOSJ_CLR	Colour attribute of *jowar suji*/*rawa*/semolina
JOSJ_OAL	Overall rating of *jowar suji*/*rawa*/semolina
JOSJ_ROS	Ease of roasting *jowar suji*/*rawa*/semolina
JOSJ_SML	Smell attribute of *jowar suji*/*rawa*/semolina
JOSJ_TX	Texture attribute of *jowar suji*/*rawa*/semolina
JOVM_A	Appearance attribute of *jowar* vermicelli
JOVM_CLR	Colour attribute of *jowar* vermicelli
JOVM_OAL	Overall rating of *jowar* vermicelli
JOVM_PRT	Preparation time of *jowar* vermicelli
JOVM_SML	Smell attribute of *jowar* vermicelli
JOVM_THK	Thickness attribute of *jowar* vermicelli
JSJ_APR	Raw appearance of *jowar suji*/*rawa*/semolina
JSJ_PRP	Preparation process of *jowar suji*/*rawa*/semolina
JSJ_TST	Taste of *jowar suji*/*rawa*/semolina
JVM_APR	Raw appearance of *jowar* vermicelli
JVM_PRP	Preparation process of *jowar* vermicelli
JVM_TST	Taste of *jowar* vermicelli
POHA_A	Appearance of cooked *jowar* flakes/*poha*
POHA_CLR	Colour of cooked *jowar* flakes/*poha*
POHA_OAL	Overall rating of cooked *jowar* flakes/*poha*
POHA_SFT	Softness of cooked *jowar poha*
POHA_SML	Smell of cooked *jowar* flakes/*poha*
POHA_STK	Stickiness/separate of cooked *jowar poha*
POHA_TST	Taste of cooked *jowar poha*
POHA_WAB	Water absorption *jowar* flakes/*poha*
SIRA_A	Appearance of (*sira*) sweet *jowar suji*/*rawa*/semolina
SIRA_CLR	Colour of (*sira*) sweet *jowar suji*/*rawa*/semolina
SIRA_OAL	Overall rating of (*sira*) sweet *jowar suji*/*rawa*/semolina
SIRA_SFT	Softness of (*sira*) sweet *jowar suji*/*rawa*/semolina
SIRA_SML	Smell of (*sira*) sweet *jowar suji*/*rawa*/semolina
SIRA_TST	Taste of (*sira*) sweet *jowar suji*/*rawa*/semolina
SIRA_TX	Texture of (*sira*) sweet *jowar suji*/*rawa*/semolina

SIRA_WTA	Water absorption of (*sira*) sweet *jowar rawa*
TE	Triennium Ending
UPIT_A	Appearance of (*uppit*) salty *jowar suji*/*rawa*/semolina
UPIT_CLR	Colour of (*uppit*) salty *jowar suji*/*rawa*/semolina
UPIT_OAL	Overall rating of (*uppit*) salty *jowar suji*/*rawa*/semolina
UPIT_SFT	Softness of (*uppit*) salty *jowar suji*/*rawa*/semolina
UPIT_SML	Smell of (*uppit*) salty *jowar suji*/*rawa*/semolina
UPIT_TST	Taste of (*uppit*) salty *jowar suji*/*rawa*/semolina
UPIT_TX	Texture of (*uppit*) salty *jowar suji*/*rawa*/semolina
UPIT_WTA	Water absorption of (*uppit*) salty *jowar suji*/*rawa*/semolina
VMSW_A	Appearance of cooked *jowar* vermicelli
VMSW_CLR	Colour of cooked *jowar* vermicelli
VMSW_OAL	Overall rating of cooked *jowar* vermicelli
VMSW_SFT	Softness of cooked *jowar* vermicelli
VMSW_SML	Smell of cooked *jowar* vermicelli
VMSW_STK	Stickiness/separate of cooked *jowar* vermicelli
VMSW_TST	Taste of cooked *jowar* vermicelli

CHAPTER 1

Sorghum: The Important Millet

Introduction

Sorghum (*Sorghum bicolor*), one of the most important staple crops grown in the semi-arid tropics of the world, is drought and heat tolerant and also resistant to climate change. It is a principal source of food for the people in Africa and Asia, while in other parts of the world, it is primarily grown for animal feed. In India, sorghum ranks fifth in production among the major cereals and is utilised as food, fodder, and industrial raw material.

Origin of Sorghum (*Jowar*)

Sorghum probably originated in North Africa about 3000 BC and was certainly being cultivated in Egypt in 2200 BC. Gradually it was spread to the tropic, semi-tropic, and arid regions of the world. It established itself in the new world since sixteenth century, though it reached China and America very recently.

Botanical Description

Sorghum is an annual plant, belonging to the family *Gramineae* or *Poaceae*. The height of the plant ranges from 0.5 m to 4.0 m. The inflorescence of sorghum is a panicle, commonly called as head. The grain is round and pointed at the base. Coloured sorghums are of white, pink, yellow, and brownish yellow colours.

Name of Sorghum in Different Languages

Sorghum is known by various names in different languages, as depicted in Table 1.1.

Table 1.1: Name of sorghum in different languages

Language	*Names*
Hindi	Jowar/Jowar chara
English	Great millet/sorghum
Kannada	Jola
Marathi	Jowari
Tamil	Cholam
Telugu	Jonnalu
Bengali	Jowar
African language	Great millet/Kafir corn/Guinea corn
Myanmar	Shallu
Thai	Kao liang
Chinese	Gaoliang
French	Sorgo
Spanish	Sorgho

Various Uses of Sorghum

Numerous types of genetically different sorghum plants are available across the world. These plants find various uses as categorised below.

- Sorgos (grown for fodder)
- Grain sorghum (grown for grain)
- Grass sorghum (wild sorghum)
- Broomcorn sorghum (used in making brooms)
- Animal feed
- Alcoholic beverages
- Bio-diesel
- Fermented and unfermented breads
- Porridge and sweets
- Extruded snacks
- Bakery items
- Sweet syrup
- Other industrial applications such as ethanol production, paper making, etc.
- Gluten-free food

Nutritional Composition of Sorghum

Sorghum, pearl millet, finger millet, and other small millets are collectively called as millets, which are now referred to as nutritious cereals. Among the millets, sorghum is the most important cereal in terms of its cropped area and use as food, though pearl millet and finger millet have also gained regional importance as staples. Sorghum plays a crucial role in rural food security in semi-arid tropics of the world.

The nutrient composition of sorghum grain indicates that it is a good source of energy, protein, vitamins, and minerals, including trace elements. Sorghum has 11.9 per cent of moisture, about 10.4 per cent of protein, and a lower fat content of 1.9 per cent. The fibre and mineral content of grain sorghum is 1.6 per cent, also being rich in dietary fibre (14.3 per cent). It provides 349 Kcal of energy and gives 72.6 g of carbohydrates (Gopalan, Rama Sastri, and Balasubramanian 1996). Starch is the major carbohydrate of the grain, and the amylose content of the starch varies from 21 to 28 per cent. The calcium, phosphorus, and iron content of sorghum is 25 mg, 222 mg, and 4.1 mg (per 100 g of edible portion), respectively (Hosmani and Chittapur 1997). Table 1.2 gives a comparison of the nutritive value of millets, rice, and maize.

Table 1.2: Nutritive value of millets in comparison with rice and maize

S. No.	*Nutrients/100 g*	*Rice*	*Bajra*	*Sorghum*	*Ragi*	*Maize*
1.	Energy (kcal)	345.0	361.0	349.00	328.0	342.0
2.	Protein (g)	6.80	11.60	10.40	7.30	11.10
3.	Fibre (g)	0.20	1.20	1.60	3.60	2.70
4.	Thiamine (mg)	0.06	0.33	0.37	0.42	0.42
5.	Riboflavin (mg)	0.06	0.25	0.13	0.19	0.10
6.	Niacin (mg)	1.90	2.30	3.10	1.10	1.80
7.	Calcium (mg)	10.0	2.00	25.0	344.0	10.0
8.	Iron (mg)	0.70	8.0	4.10	3.90	2.30
9.	Zinc (mg)	1.40	3.10	1.60	2.30	2.80

Source: Gopalan, Rama Sastri, and Balasubramanian 1996

Nutritional Importance of Sorghum

Sorghum grain has certain properties that make it suitable to be consumed by people suffering from chronic disorders. The nutritional value along with specific nutrients in the grain have been found to prevent and control lifestyle diseases and disorders.

The protein content and composition of sorghum varies from cultivar to cultivar depending on the agronomic conditions (water availability, soil fertility, temperature, and environmental conditions during grain development) and the genotype. Sorghum proteins are chiefly located in the endosperm (80 per cent), germ (16 per cent), and pericarp (3 per cent) (Taylor and Schussler 1986). Kafirins or prolamins and glutelins comprise the major protein constituents in sorghum (Warsi and Wright 1973) and are located in endosperm, while germ is rich in proteins albumins and globulins.

The carbohydrate content in sorghum comprises starch, soluble sugar, and fibre. These carbohydrates are classified into non-structural (sugars, starch, and fructosans) and structural (cellulose, hemicelluloses, and pectin substances).

Sorghum consists of considerably good amount of slow digestible starch (SDS), which has a functional property as it prolongs the digestion and absorption of carbohydrates in intestine. SDS has been found to play a major role in dietary management and in preventing metabolic disorders such as diabetes and hyperlipidemia (Asp 1994; Wursch 1997). Sorghum also contains good amount of dietary fibre in the range of 9.7–14.3 g, which acts as bulking agent and binding agent of cholesterol. It also increases transit time and retards carbohydrate absorption (Rao 2003), which helps in preventing and managing constipation, irritable bowel syndrome, and obesity.

Lipid content of sorghum grain is minor, and it is located in the scutellar area of the germ. Sorghum lipid has the potential to lower the cholesterol. A study was carried out on the effect of cholesterol absorption and plasma non-High-density Lipoprotein (HDL) cholesterol concentration in hamsters fed with sorghum lipid extract. In the study, hamsters were fed with grain sorghum lipid extract (GSL) that comprised different proportions of the diet and the results were compared with control. It was found that liver cholesterol ester concentration was significantly reduced in hamsters fed with GSL. GSL diet also lowers non-HDL cholesterol, at least in part, by inhibiting cholesterol absorption. GSL extract contains plant sterols that reduces cholesterol absorption efficiency significantly. It also contains policosanols, which inhibit endogenous cholesterol synthesis. Research findings further indicate that grain sorghum contains components that could be used as food ingredients or dietary supplements to manage cholesterol levels in humans (Carr et al. 2005).

Sorghum is a good source of minerals and vitamins that are mostly concentrated in the aleurone layer and germ. It is an important source of vitamin B except for vitamin B12 (Gazzaz et al. 1989). Yellow-coloured sorghum grain is rich in beta carotene, leutin, and zeaxanthin. However, their content varies depending on the environment and genes. Sorghum does not contain vitamin C; but it can be synthesised by soaking the grain leading to germination. Detectable amounts of other vitamins like E, K, and D are found to be present in the grain. Decortication and degermination reduce water- and fat-soluble vitamins since these micro-nutrients are located primarily in aleurone and germ portions of caryopsis. The grain is a good source of potassium and an adequate source of Mg, Fe, Zn, and Cu, but is deficient of calcium and sodium.

Sorghum is an excellent source of polyphenols, flavonoids, and condensed tannins, which are rich in antioxidants. The phenolic acids such as hydroxyl benzoic acids (galliac, vanillic, syringic, p- hydroxyl benzoic, and protocatechic acids) and hydroxycinnamic acids (coumaric, caffeic, ferulic, and sinapinic acids) are located in the pericarp, endosperm, and aleurone layers (McDonough, Rooney, and Earp 1986). Anthocyanins are the major class of flavonoids present in sorghum. Sorghum anthocyanins do not contain the hydroxyl group in the 3-position of the C-ring and are called as 3-deoxyanthocyanins. These anthocyanins are responsible for the stability of the grain at higher pH and for imparting natural colour. The 3-deoxyanthocyanins, which comprise apigeninidin (yellow) and luteolinidin (orange) in sorghum (Awika and Rooney 2004) is concentrated in highest amount in the pericarp along with flavonoids such as luteoforol and apiforol. These flavan-4-ols play an important role in mould resistance. Other flavonoids identified in sorghum grains are flavones (apigenin and luteolin), flavanones (riodictyol and eriodictyol 5-glucoside), and flavonol (kaempferol and taxifolin). Sorghum is also composed of condensed tannins, also known as proanthocyanidins or procyanidins, which are high molecular weight polyphenols (Dykes and Rooney 2006).

Health Benefits of Sorghum

As mentioned before, sorghum is a good source of energy, vitamins, and minerals, including trace elements. Apart from this, sorghum is also associated with various beneficial properties. Some of them are listed below.

- Gluten-free food: It is recommended for gluten-intolerant and celiac patients.
- Relatively low glycemic index and low glycemic load food: Such food reduces the risk of Diabetes mellitus.

- Low fat content: This lowers Low-density Lipoprotein (LDL) and improves High-density Lipoprotein (HDL) cholesterol.
- Low-calorie food: Good for people suffering from obesity, diabetes, and cardiovascular diseases (CVD).
- Rich source of antioxidants and polyphenols: These reduce oxidative stress and are anti-carcinogenic.
- Bile acid and steroid binding properties: Beneficial for obesity and CVD.
- Good source of Nitrilosides-salicylates: Fights arthritis and rheumatism.
- Rich in dietary fibre: Beneficial for people suffering from obesity, diabetes, CVD, and inflammatory bowel diseases. It also acts as a detoxifying agent.
- Rich in magnesium: Subdues depression.

Consumption of food enriched with various beneficial properties reduces the incidence of various diseases. Studies have linked sorghum in preventing and reducing the risk of the following mentioned diseases.

Celiac disease

Celiac disease (CD) is one of the most common genetic diseases, wherein genetically predisposed people display a reaction to gluten proteins found in wheat and other cereals. This disease is caused by an adverse reaction of the immune system to gluten, and it may lead to severe abdominal pains. Sorghum can be a healthy diet for those who are suffering from the disease as it is gluten free. It was reported that sorghum products could not modify the level of anti-transglutaminase antibodies even after prolonged consumption (Carolina et al. 2007).

Obesity

Obesity is an emerging lifestyle disease in India and is positively associated with several chronic diseases, including diabetes and Cardiovascular disease (CVD). Empirical evidences suggest that the intake of high dietary fibre decreases the incidence of obesity (Alfieri et al. 1995; Burkitt and Trowel 1975). Food rich in dietary fibre improves the large bowel function and slows down the digestion and absorption process, thereby reducing the risk of chronic diseases (Ali et al. 1982; Schneeman and Tietyen 1994). Sorghum is rich in dietary fibre and has unique chemical and physical characteristics (in providing bulk to the diet; viscosity; and water holding and absorption capacity) which determine the subsequent physiological behaviour. It satisfies hunger pangs, increases satiety, and thereby, reduces the risk of the onset of obesity.

Diabetes

Diabetes mellitus, a complex metabolic disease is a major health concern in many countries. The prevalence of type 2 diabetes (T2D) is rising across the world at an alarming rate, especially in India, where 14.3 per cent of the population is afflicted with this disease. Prevention of T2D will require measures that promote healthy diet and lifestyle, which include having balanced diet and physical activity. People who eat three or more servings of whole grains in a day, especially that of high-fibre cereals, are less likely to develop insulin resistance and metabolic syndrome, common precursors of both T2D and Coronary Heart Disease (CHD) (Mckeown et al. 2004). Diet rich in whole grain food tends to decrease LDL cholesterol, triglycerides, and blood pressure and increases HDL cholesterol (Anderson 2003). Sorghum is rich in dietary fibre and has low glycemic index, which could help in the prevention and control of T2D in Indians. The presence of fibre, magnesium, vitamin E, phenolic compounds, and tannins in food reduces the risk of diabetes as these components slow down the incidence of sudden increase in blood glucose and insulin levels (Montonen et al. 2003).

National Institute of Nutrition (ICMR) in 2010 assessed the Glycemic Index (GI) of sorghum-based food in collaboration with the Directorate of Sorghum Research (DSR), Hyderabad, under the National Agricultural Innovation Project (NAIP). The results revealed that sorghum-based food has low GI, and it reduces the postprandial blood glucose level and glycosilated hemoglobin. Another study also points to the fact that blood glucose level of non-obese patients with non-insulin-dependent diabetes mellitus (NIDDM), who consumed sorghum bran *papadi*, showed considerable reduction (Shinde 2004).

Coronary heart disease

Empirical evidences (Slavin 2003; Slavin and Slavin 2004) suggest that the regular consumption of whole grains reduces the risk of CVD (Anderson and Hanna 1999). A long-term prospective study by Heidemann et al. (2008) showed that the regular consumption of balanced diet, including whole grains, vegetables, fruits, fish, and poultry products reduces the risk of CVD and total mortality. In 2004, Harvard professionals (Jensen et al. 2004) analysed the diet records of 27,000 men aged between 40 and 75 years for a period of 14 years and found that the intake of 40 g of whole grains per day reduces the risk of CHD by 20 per cent.

Cancer

Anti-carcinogenic properties of sorghum have been well documented. *In vivo* and *in vitro* studies have shown that the consumption of sorghum has positive health impacts on people suffering from cancer. The polyphenols and tannins present in sorghum have anti-mutagenic and anti-carcinogenic properties (Grimmer, Parbhoo, and McGrath 1992) and can act against human melanoma cells as well as positive melanogenic activity (Gomez-Cordoves et al. 2001). Parbhoo et al. (1995) reported that in the rat liver, procyanidin extracts may induce cytochrome P-450, a protein that is capable of converting certain promutagens to mutagenic derivatives. Epidemiological data from Sachxi Province, China, and from different parts of the world (Van Rensburg 1981) showed that the incidence of oesophageal cancer was low in population that consumed sorghum. In each country, the authors studied 21 communities over a period of six years and found that consumption of sorghum led to lower mortality from oesophageal cancer as compared to that in wheat and corn consumption. It can be concluded that anti-carcinogenic compounds present in sorghum lowered the risk of oesophageal cancer.

Oxidative stress

Free radicals are atoms or molecules with unpaired electrons and can be formed by interaction with other molecules. These free radicals are responsible for oxidative stress related pathogenesis of various diseases such as Alzheimer's disease, myocardial infarction, atherosclerosis, Parkinson disease, auto-immune diseases, etc. Antioxidants play an important role in preventing oxidation process, thereby reducing cellular damage. Sorghum has potential antioxidant capacity, which can act against reactive oxidative species.

Sorghum Consumption Pattern

Sorghum is a coarse grain, primarily used as food in semi-arid tropics, and as livestock feed and industrial raw material in developed countries. Earlier, it was used to be a traditional subsistence crop, but now it is gradually emerging as a commercial crop. Globally, the use of sorghum for feed is the main driving force for its production and international trade. Sweet sorghum is used for the production of ethanol and jaggery, and making paper.

Various Uses of Sorghum Across the World

In South Africa, sorghum meal is used for the preparation of porridge, sweets, and savouries. In the United States, syrup prepared from sorghum is used as an alternative to sweetener for the preparation of bakery items such as biscuits and cakes.

In Middle East countries, sorghum grain is used for the preparation of soups, porridges, cakes, etc. It is also used for the preparation of bread in combination with other flours. In China, it is the main ingredient for the production of alcohol. Sorghum crop is used as cattle feed, and its stems can be used for wall-board making and biodegradable packaging.

Uses of Sorghum in India

Bhakri (*jolada roti* in northern Karnataka), a variety of unleavened bread usually made from sorghum, is the staple diet in many parts of India, such as Maharashtra and northern Karnataka. In eastern Karnataka and Rayalaseema, Andhra Pradesh, *roti* (*jonna rotte*) made with sorghum is a staple food.

Sorghum has a distinct advantage, when compared to other major fine cereals, of being drought-resistant, and many subsistence farmers in various regions cultivate sorghum as a staple food crop for consumption at home (Murty and Kumar 1995). Traditional recipes of sorghum along with broom corn were incorporated into American cuisine by the nineteenth century, and it was also widely used to brew beers.

Sorghum and millets are nutritionally superior to other cereals, but their low demand in cereal basket may be due to the factors like urbanisation, rising incomes, lifestyle changes, social status attached to fine grains, subsidised supply of fine grains, low productivity of coarse cereals, short shelf-life of flour, and more effort needed in making *rotis* (unleavened bread). The government should consider including sorghum in public distribution system (PDS). Table 1.3 indicates the utilisation of sorghum and other food grains in food and for other uses.

Table 1.3: Utilisation of sorghum and other food grains in direct consumption and other uses (million tonnes)

Commodity	*Food*		*Other Uses*		*Total*	
	2003/04	*2011/12*	*2003/04*	*2011/12*	*2003/04*	*2011/12*
Rice	81.6	92.5	7.3	13.9	88.9	106.3
Wheat	56.7	62.5	16.4	17.4	73.1	79.9
Sorghum	5.4	4.2	1.8	7.2	7.2	11.4
Bajra	4.7	5.0	4.0	6.9	8.7	11.8
Maize	2.8	2.5	11.6	20.3	14.4	22.8
Small millets	1.7	1.3	0.3	0.8	2.0	2.1
Total cereals	153.5	168.6	41.3	66.4	194.8	235.0
Total food grain	165.4	183.5	45.8	71.4	211.3	254.9

Source: National Centre for Agricultural Economics and Policy Research (NCAP) 2007

Roti (unleavened bread), *sankati, annam, pops, and ganji* (thin porridge) were a part of the most popular sorghum recipes in rural India. The traditional recipes to make them are tedious and time consuming. The major constraint is that the flour becomes rancid within a few days after milling. To overcome this problem, sorghum grains are subjected to processing treatments like malting, popping, and dry heat treatment before developing the final product, which leads to a sorghum product with longer shelf-life and better nutritional profile.

Opportunity of Developing Value-added Sorghum Products

Over the years, there has been a drastic change in farming, production, and consumption of sorghum in rural and urban India. It is termed as the poor man's food, and hence the effect is more visible in rural areas. Generally, the demand for such crops is higher where the farming is cost-effective, the product is highly marketable with long shelf life, and is readily processed, creating employment. As *jowar* has no supporting government policies (including no subsidies as in rice and wheat, and not included in PDS, etc.), farmers have little interest in its cultivation. Finding market for the cultivated grain is so difficult for them that they divert to other cereals and cash crops. The grains cannot be stored for a long time, and though the grains of *rabi* season are of good quality but the yields are very low and the *kharif* sorghum gives high yields but the quality is not suitable for consumption. Primary and secondary processing of sorghum is fraught with difficulties, which makes the rural and urban population shift to other cereals. These are observed to be the main reasons for the decline in the cultivation of sorghum. The shift to other food by urban population can be addressed by innovation in secondary processing methods. DSR has come up with excellent ready-to-cook and ready-to-eat sorghum value-added products, which have created a great demand in the urban markets where there is a growing concern for health-related problems. Some of the products commercialised by DSR are sorghum-rich multigrain *atta*, sorghum *rawa* (fine and coarse rawa for *upma, dosa, idli*, etc.), extruded products (vermicelli and pasta), flaking (breakfast cereal), baked products (biscuits and cookies), and instant mixes (*dosa* mix, *idli* mix, *peda* mix, etc).

CHAPTER 2

Availability and Quality of Sorghum

Sorghum Cultivation Areas

Sorghum is a tropical grain grown primarily in semi-arid parts of the world. In Africa, a major growing area runs across West Africa south of the Sahara, through Sudan, Ethiopia, and Somalia. It is grown in upper Egypt, Uganda, Kenya, Tanzania, Burundi, and Zambia. It is an important crop in India, Pakistan, Thailand, central and northern China, Australia, the drier areas of Argentina and Brazil, Venezuela, USA, France, and Italy. The crop is spread over the drier areas of the world; it grows better when the weather is dry and cool. Sorghum is a staple food for about 300 million people worldwide.

Production of Sorghum Worldwide

Globally, sorghum is cultivated in an area of 42–43 million hectares leading to a production of 59-60 million tonnes of crop. In the last 18 years (1994-2011), average production of sorghum in the world is pegged at 59–60 million tonnes. The compounded annual growth rate (CAGR) in global production of sorghum has slightly declined by -0.5 per cent. The global production trend of sorghum is given in Fig. 2.1.

Country wise production of sorghum

United States is the largest producer of sorghum in the world. Annual average production of sorghum in the country during the year 1994–2011 was about 11.9-12.0 million tonnes. The compounded growth during these periods was -4.3 per cent annually. India is the second largest producer of sorghum in the world; on an average, the country produces around 7.5–7.8 million tonnes of sorghum annually. Nigeria ranks third and Mexico fourth in terms of the production of sorghum globally as the average production of sorghum during the year 1994–2011 in these countries was 7.68 and 5.94 million tonnes, respectively. Table 2.1 gives the production of sorghum in various countries globally.

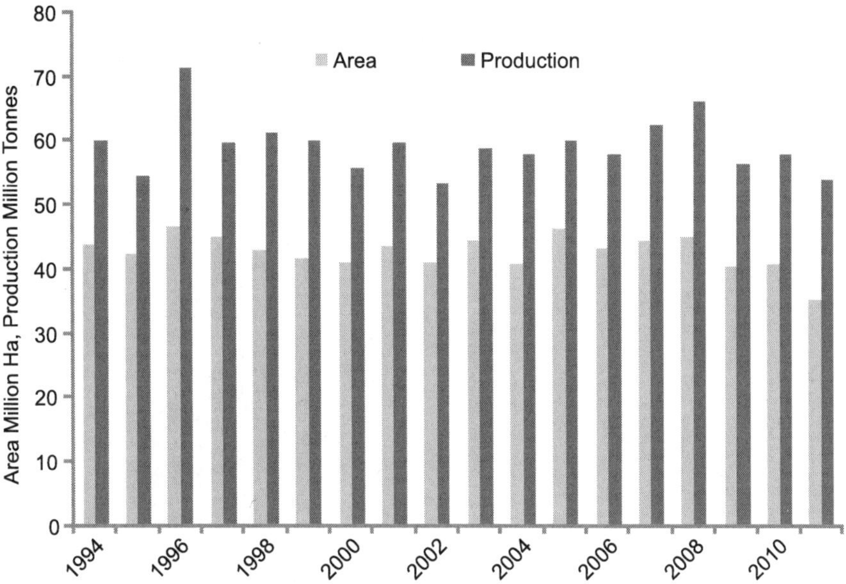

Fig. 2.1: Global production trend of sorghum

Source: FAO Statistics (URL: http://faostat.fao.org/site/567/default.aspx#ancor)

Table 2.1: Global production of sorghum, by country

	Average Production (1994–2011)	*Share in World Production (%)*	*Growth (%)*
United States of America	11.91	20	-4.3
India	7.81	13.1	-1.7
Nigeria	7.68	12.9	0.6
Mexico	5.94	10	1.7
Sudan (former)	3.47	5.8	0.2
China	3.13	5.3	-6.8
Argentina	2.77	4.7	1.5
Ethiopia	1.92	3.2	6.9
Australia	1.84	3.1	3.4
Burkina Faso	1.39	2.3	3.2
Brazil	1.16	2	12.5
Egypt	0.83	1.4	0.6
Mali	0.79	1.3	4.4
Chad	0.53	0.9	3.1
World	**59.42**		**-0.3**

Source: FAO Statistics (URL: http://faostat.fao.org/site/567/default.aspx#ancor)

The top eight countries account for around 75 per cent of the global production of sorghum. The major share is contributed by United States (20 per cent), followed by India (13.1 per cent), Nigeria (12.9 per cent), Mexico (10 per cent), Sudan (5.8 per cent), China (5.3 per cent), and Argentina (4.7 per cent). The global average production of sorghum during the period 1994–2011 was 59.42 million tonnes. The share of various countries in sorghum production globally is summarised in Fig. 2.2.

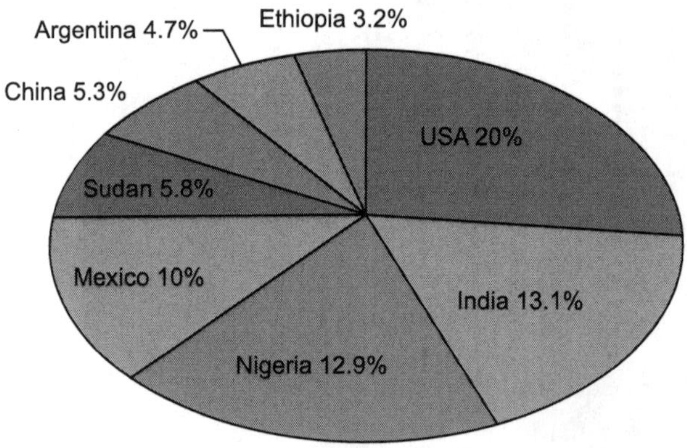

Fig. 2.2: Percentage share of countries in global sorghum production (1994–2011)
Source: FAO Statistics (URL: http://faostat.fao.org/site/567/default.aspx#ancor)

Current Status of Sorghum in India

Sorghum is the third important cereal in India after rice and wheat. It is cultivated mainly in two seasons, namely, *kharif* and *rabi*. *Kharif* sorghum is characterised by hybrids producing higher yields, while *rabi* sorghum is predominantly characterised by improved varieties.

Production of Sorghum

In the last 20 years (1993–2012), India, on an average, produced around 8.0 million tonnes of sorghum in an area of 9.4 million hectare. Average productivity during the period was 900 kg/ha. There is a minor decline in the CAGR of area (–3.5 per cent) and production (–2 per cent) during the same period. The yield or productivity of sorghum has increased substantially. Average yield during the last 10 years was 874 kg/ha, and in the last five years, average yield recorded was around 950 kg/ha. Fig. 2.3 gives the area under sorghum cultivation along with its production and yield.

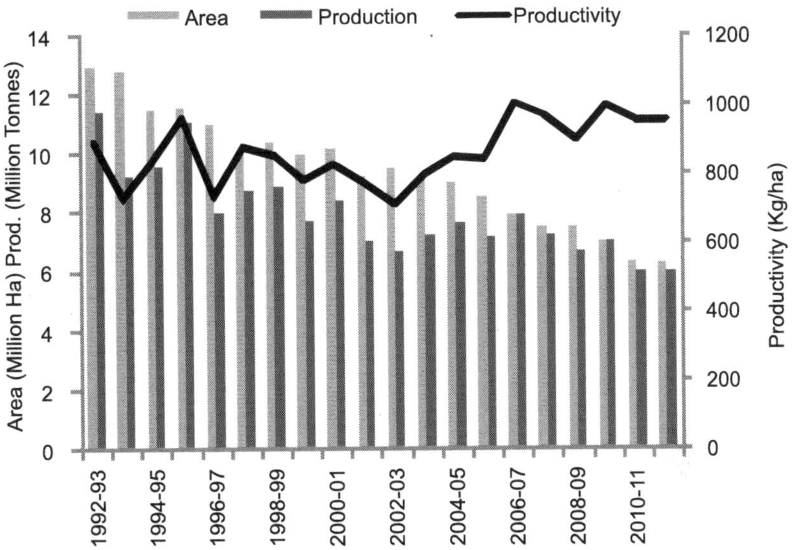

Fig. 2.3: Average area under cultivation, production, and yield of sorghum in India

Source: FAO Statistics (URL: http://faostat.fao.org/site/339/default.aspx)

It is important to consider that though the area under sorghum cultivation is declining from year to year, the production levels have not declined drastically owing to the adoption of improved cultivars. The CAGR in yield during 1993–2011 was 1 per cent, and during the last 10 years (2002–11), it was 3 per cent.

State wise production of sorghum

Maharashtra is the largest producer of sorghum in India. The state occupies almost 35 per cent of the total cultivated area and accounts for 41.5 per cent of the total production of sorghum in the country. Karnataka and Madhya Pradesh are second and third largest producers in the country, respectively. These three states together contribute around 62 per cent in the total production. Andhra Pradesh, Rajasthan, and Tamil Nadu are the other major sorghum producing states in India. The contribution of various states to sorghum production is given in Fig. 2.4.

Productivity of sorghum in Rajasthan is the lowest, mainly because it is grown under rain-fed conditions, and the residue moisture in the soil is lesser as compared to the other sorghum producing states. Fig. 2.5 gives the area under sorghum cultivation in various states of India.

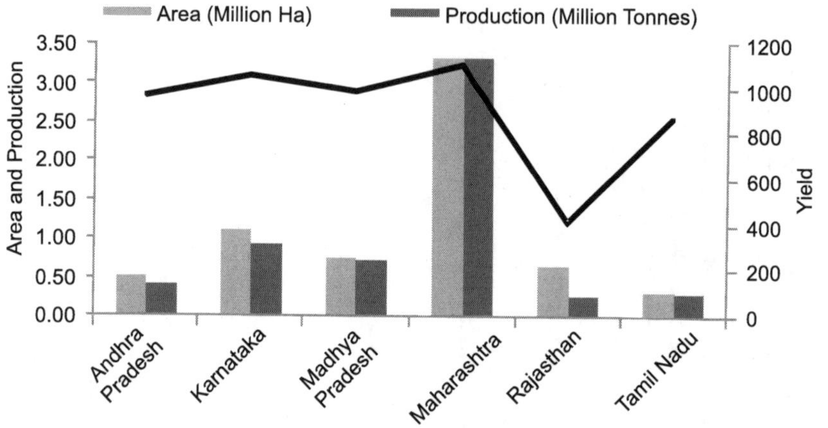

Fig. 2.4: Average area under cultivation, production, and yield of sorghum, by state (1993-2011)

Source: Directorate of Economics and Statistics (DAC), Ministry of Agriculture, Govt of India (URL: http://eands.dacnet.nic.in/)

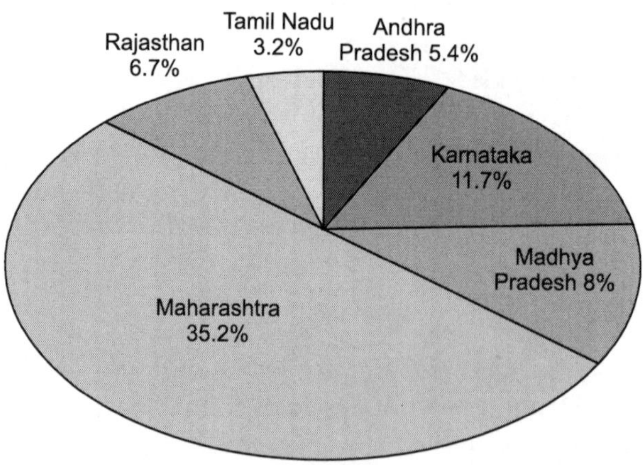

Fig. 2.5: Percentage contribution of major sorghum cultivating states in total average area under cultivation (1993–2011)

Source: Directorate of Economics and Statistics (DAC), Ministry of Agriculture, Govt of India (URL: http://eands.dacnet.nic.in/)

Total average area under sorghum cultivation in the country during 1993–2011 was 9.41 million hectares, while total average production of sorghum in the country during the same period was 7.98 million tonnes. Fig. 2.6 depicts the contribution of major sorghum cultivating states in total average production.

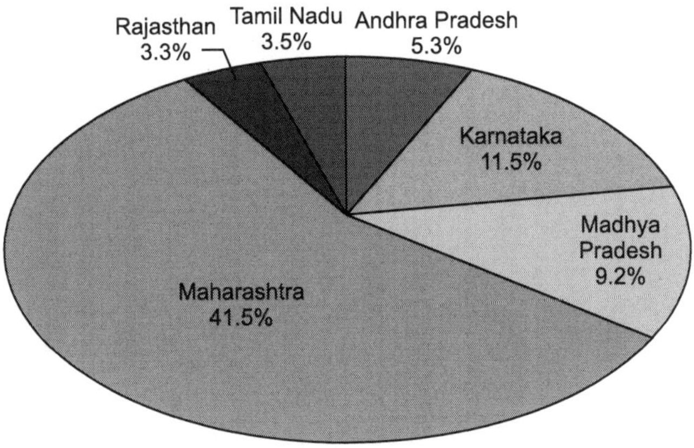

Fig. 2.6: Percentage contribution of major sorghum cultivating states in total average production (1993-2011)

Source: Directorate of Economics and Statistics (DAC), Ministry of Agriculture, Govt of India (URL: http://eands.dacnet.nic.in/)

Season wise production of sorghum

The percentage share of sorghum production during *kharif* and *rabi* seasons in the last three years is 45 per cent and 55 per cent, respectively. Madhya Pradesh and Rajasthan are the leading producers of *kharif* sorghum. Area under sorghum cultivation and its production during the *rabi* season is almost negligible in both the states. Fig. 2.7 shows the production of sorghum in *kharif* and *rabi* seasons in the country.

Quality of Sorghum

Though the *kharif* sorghum cultivars give higher grain and fodder yields, they are prone to mould due to excess or untimely rain at the time of harvest, usually during October. This rain affects the grain quality and makes it unfit for human consumption. Thus, the *kharif* grain fetches lower price due to its poor quality and is consequently increasingly diverted for non-food industrial uses in almost all the years (i.e., atleast thrice in every five years).

Current trend of *kharif* sorghum is mainly determined to fulfil the fodder requirement from the crop. Introduction of hybrids witnessed a major change in the Indian sorghum farming, specifically in *kharif* season.

Though the productivity of *rabi* sorghum is comparatively lower than the rain-dependent *kharif* grain, the former is far more superior in quality, as it

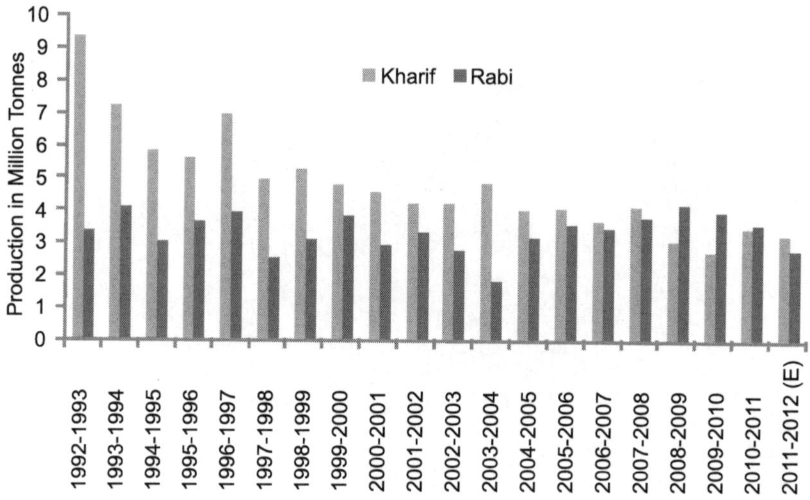

Fig. 2.7: Production of sorghum in India, by season

Source: Directorate of Economics and Statistics (DAC), Ministry of Agriculture, Govt of India (URL: http://eands.dacnet.nic.in/)

matures in cold climate. However, of late, in some areas, the *rabi* sorghum grain is cultivated with the help of irrigation, thus yielding almost double the quantity. Recently released *rabi* varieties, namely, *Phule* series, CSV 14R, and CSV 18, have also been received well by farmers. Due to its better quality, the profit realised by *rabi* grain is almost double of that of *kharif* grain.

In recent years, machinery/technologies are made available for dehulling. These prototypes are used by R&D institutes and small-scale entrepreneurs. Despite the fact that the dehuller decorticates the outer layer and may lead to partial loss of minerals and vitamins, such grains can be used for direct food consumption. These can also be used for making other sorghum-based products like RTC/RTE products, which are easy to make, taste good and have nutritive value.

The market price of de-hulled grains is higher than the moulded grains. Sorghum is fast emerging as a commercial crop and shows a remarkable potential for various industrial uses. Its use as cattle feed, poultry feed, and in brewing potable alcohol is already well known. Thus, *kharif* sorghum grain can be primarily utilised in poultry, brewing alcohol, and other industrial sectors, apart from it being used as a direct staple food.

CHAPTER 3

Marketing of Sorghum

Marketing of Sorghum in India

The marketing process of sorghum in India is generally smooth and constraint-free as it is relatively easy to market it as a food grain. At times, linked or tied output and credit markets lead to distress sale by small and marginal farmers. The existing marketing arrangements may not be favourable for the industries that use sorghum and millet. These industries may resort to bypassing the regular channels to obtain sorghum and millet more easily for dual purpose: for food and for industrial applications (Marsland and Rao 1999). With the gradual commercialisation of Indian agriculture and its integration with the world markets coupled with the increasing demand for sorghum, marketing of sorghum, an erstwhile subsistence crop, assumes great importance.

Although significant quantity of sorghum is traded, markets are still not well developed. There is a need to popularise value-added and ready-to-use sorghum food products. It is important that there is a collaboration with different industries and a general understanding of the behaviour of market players. New channels for the promotion of alternate uses of sorghum like in syrup, industrial products, jaggery from sweet sorghum, brewing of potable alcohol, and fuel alcohol from sweet stalks need to be explored.

Marketing of sorghum grain fits no ordinary marketing model. Traditionally, only 8–9 per cent of the crop carries over to the next marketing year. There are two primary reasons for this low carry-over. First, sorghum as a commodity moves in and out of feeding channels as a replacement for maize in most markets due to its lower price. Second, the traditional market for sorghum is a replacement of starch market. Starch is marketed as binders for making wallboard and for ethanol production. One of the fastest growing areas of utilization of sorghum is the production of ethanol, which now uses approximately 12 per cent of the total sorghum grain production in the country.

What is Marketable Surplus?

The marketable surplus is the surpluses of produce that exists after deducting the expenses of labour, tools, fertiliser, and land. Even the crop for self-consumption and the use of seeds for the next year are also accounted for before calculating the marketable surplus. In this way, marketable surplus is what a farmer makes for his personal labour; the portion of the harvest that a farmer can sell in the market to earn profit. The concepts of marketable and marketed surplus are being considered as crucial factors in the process of economic development. Marketable surplus represents the theoretical surplus available for disposal with which the producer is left after his genuine requirements of the family consumption, and payment of wages in kind, feed, and seed have been met. It is the proportion of the total output that should normally be brought for sales.

All farmers with different sizes of holdings may not be expected necessarily to have marketable surplus. It depends on the distribution of grain among various components of retentions. If they consume more at farm level itself, they may have a little or negligible size of surplus. If they consume less, retaining sufficient amount of food grains for domestic consumption, they may be expected to have more marketable surplus. When farmers sell more than their potential marketable surplus, it may be termed as forced sales or distress sales.

The farm households resort to sale of *jowar* over a period of time, depending upon the needs. The small category farmers, who are hard pressed to meet the financial obligations, resort to, distress sales, immediately after harvest, whereas the farmers with better holding capacity dispose off the produce when the market is favourable. The total produce sold by the farm households is termed as the marketed surplus, i.e., the surplus of produce, which is sold in the market irrespective of the actual requirement of the farm households. This is the quantity that is made available for consumption to the non-farming population, including the buy-back by the farm households (AGMARKNET 2002).

Marketable Surplus of Sorghum in India

Normally sorghum is mainly cultivated by resource-poor farmers to meet household requirements of food and fodder and a small surplus for the market. After harvest, sorghum grain may be retained on farm, to be used for home consumption or as seed or for payment in kind. The surplus grain is sold either in a regulated market through commission agents in the markets or through a broker (middleman) at the village. In general, marketable surplus of sorghum is not huge, but because it is one of the main staple foods, it is reported that the movement of sorghum from surplus areas to deficit areas and between rural markets within each state is common. Sorghum was found in all the rural and urban markets. However, it was observed that sorghum supply is very limited in all the rural markets as well as in the urban markets.

Marketable surplus declined over the last 10 years (except in 2003/04) owing to the steady decline in feed consumption demand with shift towards other crops like maize, bajra taking price advantage (Dayakar Rao et al. 2010a). As a result, of the total Indian sorghum production, an average 40 per cent is available as marketable surplus. However, this percentage of diversions kept fluctuating depending upon the production in a given year. Further, majority of *kharif* output is used for animal feed owing to the poor quality of grains and sizeable acreage under hybrid crop; the human consumption of the same is lower and there is an increasing dependence on *rabi* output for human consumption. Fig. 3.1 gives the marketable surplus of sorghum in India.

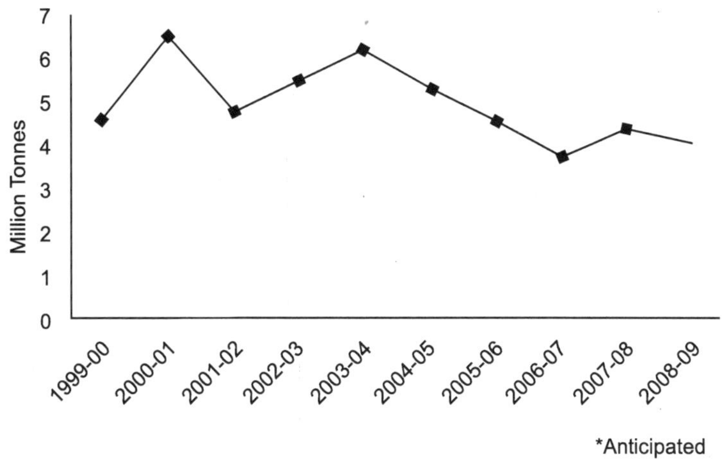

Fig. 3.1: Marketable surplus of sorghum in India

Price Trend of Sorghum

At the global level, sorghum is almost exclusively traded as feed grain, and its market price is closely related to the production and trade of other food grains such as maize, wheat, and barley.

Seasonality in International Price of Sorghum

The seasonality in international price of sorghum is depicted in Fig. 3.2.

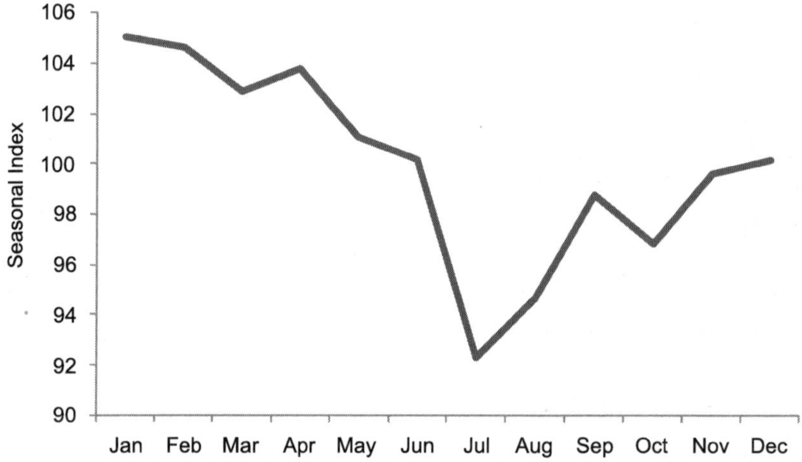

Fig. 3.2: Seasonality in monthly price of sorghum in international market (2006-11)

Source: Index Mundi (URL: http://www.indexmundi.com/commodities/?commodity = sorghum)

The price of sorghum starts declining from the month of April and reaches the lowest level during the month of July. The prices start increasing and remain high upto the month of March. During the months of January and February the prices of sorghum reach the peak level. The monthly average price trend of sorghum in international market is shown in Fig. 3.3.

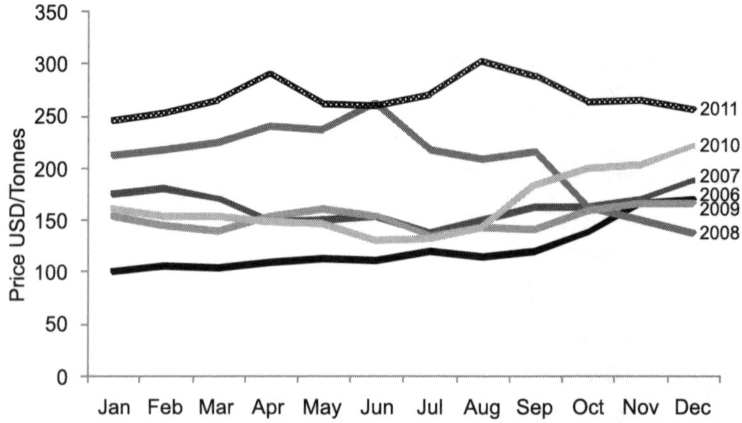

Fig. 3.3: Monthly average price trend of sorghum in international market

Source: Index Mundi (URL: http://www.indexmundi.com/commodities/?commodity = sorghum)

Except 2008 and 2011, sorghum prices were range bound. The average price of sorghum during 2008 and 2011 was USD 208 and USD 268 per tonne, respectively.

Sorghum Price Scenario in India

Index of prices of sorghum in India declined sharply until 1990s and rose above the inflation since 1995, perhaps indicating renewed demand for sorghum particularly for non-food uses. Sorghum prices are generally more volatile (large year-to-year fluctuations) as compared to rice and wheat prices since the production of sorghum is dependent on rainfall. Over the last four decades, the real prices of sorghum have exhibited a cyclical pattern, showing a decreasing trend.

The monthly average price of white sorghum for six years (2007-12) in Bengaluru market has been analysed and depicted in Fig. 3.4. The price of sorghum in Indian market has seen an upward trend since 2011. Average sorghum price during 2011 and 2012 was Rs 2005/quintal and Rs 2265/quintal, respectively. The highest price of sorghum in both the year was Rs 2500/quintal during the month of September.

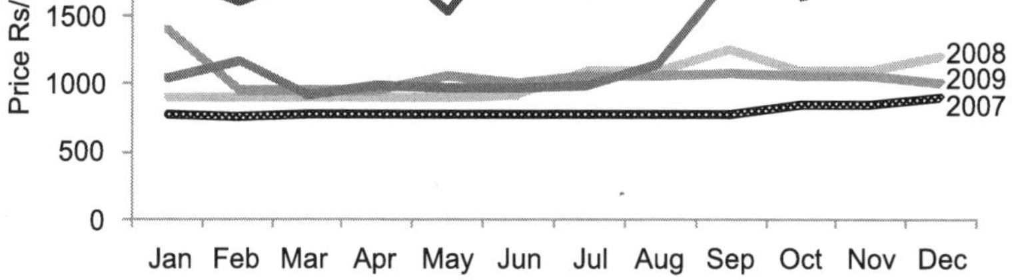

Fig. 3.4: Monthly average price trend of sorghum in Bengaluru, India

Source: Agmarknet (URL: www.agmarknet.nic.in)

Seasonality in Price of Sorghum in India

Seasonality in the price of sorghum appears to be similar to the international market price seasonality. The price of white sorghum is lowest during the month of July and then it starts moving up. During the months of September, January, and March, sorghum price reaches the highest level (Fig. 3.5).

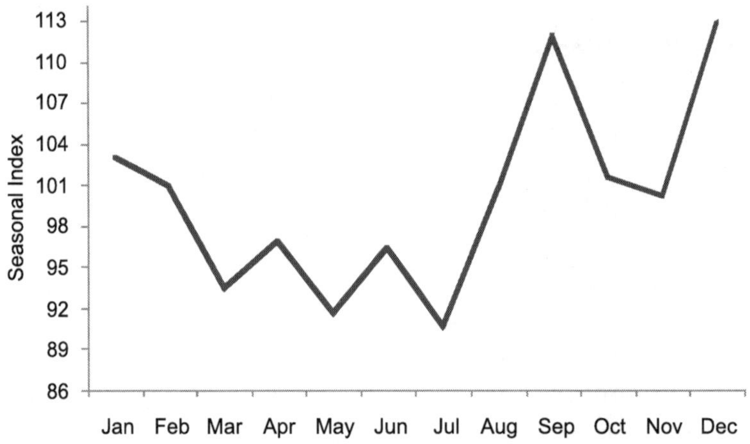

Fig. 3.5: Seasonality in monthly price of sorghum in Bengaluru, India

Source: Agmarknet (URL: www.agmarknet.nic.in)

Minimum Support Price for Sorghum

Minimum support price (MSP) announced by the government plays an instrumental role in influencing the decision of the farming community with respect to the type of crop they cultivate. In the case of sorghum too this holds true. In the last four years, sorghum prices have gone up, partly on account of the steady increase in demand and increase in MSP paid to farmers by the government procurement agencies, which determines the open market prices to a significant extent (Table 3.1). Thus, during 2012/13, the MSP for *jowar* has been raised by almost 53 per cent for hybrids and 52 per cent for *Maldandi.*

The government has included the *Maldandi* variety of sorghum in the MSP from 2006–07.

However, higher price realisation vis-à-vis MSP with robust export demand in the last couple of years and drought conditions in Australia led to export opportunity to countries like Brazil and Japan, which played a crucial role in determining acreages in the recent past. For the current year, Indian government hiked sorghum MSP by 40 per cent and the same encouraged the farmers to go for higher acreages during the current crop season. Considering the current market sentiments, probability is high that market price would not test MSP because of lower acreage and strong demand from industries.

Table 3.1: Minimum support price (MSP) of sorghum, by year

Year	*Hybrid (Rs/quintal)*	*Maldandi (Rs/quintal)*
2000/01	445	N.A.
2001/02	485	N.A.
2002/03	485	N.A.
2003/04	505	N.A.
2004/05	515	N.A.
2005/06	525	N.A.
2006/07	540	555
2007/08	600	620
2008/09	840	860
2009/10	840	860
2010/11	880	900
2011/12	980	1000
2012/13	1500	1520
Increase in MSP 2012/13 over 2000/01	1055 (237%)	N.A.
Increase in MSP 2010/11 over 2009/10	40 (4.8%)	40 (4.7%)
Increase in MSP 2011/12 over 2010/11	100 (11.4%)	100 (11.1%)
Increase in MSP 2012/13 over 2011/12	520 (53.0%)	520 (52.0%)

Segments of Sorghum Consumption

Sorghum is the staple food in central and western regions of Maharashtra and the northern regions of Karnataka and Andhra Pradesh. Per capita food consumption of sorghum in rural producing areas is more stable, and is usually considerably higher than that in the urban centres, and within these rural areas, consumption tends to be highest in the poorest, most food-insecure regions. Between 1972/73 and 2004/05, the annual per capita consumption of sorghum at the all-India level declined sharply by 68 per cent (from 8.5 kg to 2.7 kg) in urban areas and by 70 per cent (from 19.1 kg to 5.2 kg) in rural areas of India (Basavaraj and Rao 2011). During 1999–2000, sorghum consumption in India stood at 8.85 million tonnes with human consumption accounting for 77 per cent, but it declined to 7.70 million tonnes with human consumption accounting for 6.27 million tonnes and the remaining used as poultry and dairy feed (Dayakar Rao et al. 2010c). This decline in consumption of sorghum both in rural and urban areas at all-India level is due to several factors but the most important ones are increase in per capita income, growing urbanisation, and changing tastes and preferences (Chand 2007).

The low-income consumers (about 30 per cent of population in rural and urban areas each) account for 35 per cent and 49 per cent of sorghum consumption in rural and urban areas of India, respectively. Their per capita consumption is also the highest among the three income groups (Table 3.2). The share of high-income consumers is low both in rural and particularly in urban India. However, in inland regions of central, eastern, and western Maharashtra and northern Karnataka, sorghum is still an important crop with annual per capita consumption in rural areas ranging from 31.8 kg to 54.2 kg and in urban areas from 9.9 kg to 34.0 kg. Moreover, in the rural parts of these areas as well as northern Maharashtra, the inverse relationship between sorghum consumption and income is less apparent than elsewhere. Both states grow a large proportion of the Indian crop (Maharashtra 47 per cent and Karnataka 20 per cent) with lesser amounts being grown in Andhra Pradesh (9 per cent). There is variation within regions of the three states with respect to the total amounts grown and the proportions of the two sorghum types, *rabi* and *kharif*. The former is preferred as food, as the latter tends to be of poorer quality and subject to grain mould: it is consequently mostly used as a feed ingredient in the poultry and livestock industries and as raw material for the alcohol industry. However, there are considerable shortfalls of *rabi* sorghum in most of the regions of the three states with the deficit for human consumption being made up from better quality samples of *kharif* sorghum, which is less expensive than the *rabi* type. It is suggested that sorghum should be included in the public distribution system (PDS) in Maharashtra, Karnataka, and Andhra Pradesh, and the research should be devoted to increasing the productivity of the *rabi* type, in order to make it affordable for the poorer sections of communities.

Table 3.2: Consumption of sorghum in India by different income class groups

Expenditure Category	*Consumption*		*Per capita Consumption (kg/year)*	*Population (%)*
	('000 t)	*(%)*		
Rural average consumption				
Low (less than Rs 365 per month	1421.4	35.3	6.09	30.2
Medium (Rs 365-890 per month)	2346.4	58.2	5.09	59.8
High (greater than Rs 890 per month)	263.7	6.5	3.42	10.0
Urban average consumption				
Low (less than Rs 580 per month	416.4	49.3	4.46	30.2
Medium (Rs 580-1880 per month)	404.6	47.9	2.19	59.8
High (greater than Rs 1880 per month)	24.4	2.9	0.79	10.0

Source: Compiled from 'Level and pattern of consumer expenditure', National Sample Survey Office (NSSO), 61 Round, 2004–05

CHAPTER 4

Development of Convenient Foods through Diversified Processing Technologies

Need for Product Development

Food processing industry plays a vital role in the economic growth of the country. It helps in increasing the production and income of farmers, reducing the post-harvest losses, generating employment, and creating market for agricultural products. Food processing is defined as a process of modifying or adding value to the agricultural products by various methods. The main objective of product development is to develop food products with different characteristics, which gives additional benefits to the consumers. Products that are developed with innovative processing technologies improve the shelf life, quality, palatability, and nutritional composition of food.

In India, food processing industry has emerged as a key component in the manufacturing sector. Previously, food processing was majorly confined to the food preservation, packaging, and transportation. Emerging processing technologies have made it easier to produce ready-to-eat (RTE) and ready-to-cook (RTC) food. Current government policies as well as economic growth of the country have contributed to the growth of processing industries, development of various food products, and expansion of the market. The demand for processed and convenience food is increasing constantly because of the rise in income levels, urbanisation, steadily growing middle class and younger generation, changing lifestyle and food habits of the people, rising export opportunities, and proximity to key destinations. Among the food processing industries, grain processing industry is the key component of the food sector, which shares about 40 per cent of the market.

Over the past decade, sorghum has attained more importance for human consumption due to its nutritional and health benefits. India is one of the primary or secondary centres of diversity for sorghum and other several millet crops. The cost of cultivation of sorghum is also less as compared to other fine cereals. However, sorghum utilisation was limited in processing sector, and its consumption declined due to urbanisation, previous government policies, lack of processing technologies and awareness on its health and nutritional benefits, tedious and time-consuming process involved in preparation of food out of it, and non-availability of convenient foods. These drawbacks resulted in a marked shift towards more convenient/processed products of fine cereals like rice and wheat. In some parts of India, this trend has led to increase in malnourishment among population in low economic strata.

Food and nutritional security is the major challenge in developing countries. In India, more than one-third of the population has been estimated to be poor and of that, half is malnourished. This conflicts the assurance of food security in the country. Rice and wheat alone cannot suffice in meeting the food and nutritional security in the wake of increasing population. Cereals and millets should be considered a dependable food source due to their rich nutritional profile. However, efforts involved in processing millets are highly inadequate, which comes in the way of their being accepted as convenient food.

Sorghum has unique nutritional properties due to which it shows a lot of promise in addressing the nutritional security of our country. Popularising these nutritional benefits coupled with innovative processing interventions such as milling, baking, flaking, and extrusion cooking is expected to raise the demand for sorghum and would lead to its increased utilisation. Further, such interventions will result in the development of RTE and RTC food, which would find popularity with the working middle class.

Sorghum can be milled to produce starch and grits (semolina), which form the base of many Indian traditional recipes. As the primary and secondary millet processing methods are not widely followed, the resulting products do not have wide utilisation. Study needs to be done to find out whether the sorghum products fulfil the consumption criteria and whether these could be consumed on a daily basis in place of other cereals. Also, appropriate processing technologies should be used to increase the shelf-life of the sorghum products and their palatability.

Development of RTC/RTE Food

Processing of sorghum ensures the availability of sorghum products throughout the year and across the country in a safe and convenient form. RTE and RTC food is semi-cooked or processed food. Under the National Agricultural

Innovation Project, Directorate of Sorghum Research (DSR)-led consortium developed novel processing methods by standardising sorghum product technologies, optimising the conditions for each processing technology, and retrofitting the machinery for processing. DSR has developed 18 products, but has chosen only six products that have been found to have potential market based on the findings from a market study conducted by an independent agency. These products were labelled and packed professionally and launched on a pilot scale in retail stores in Hyderabad under *Eatrite* brand specifically registered by DSR. These products were multigrain flour, *suji* (three variants), RTC (flakes, pasta, and vermicelli), and RTE products (roasted flakes and biscuits).

DSR Innovations in Sorghum Processing

The processing interventions of DSR-led consortium are aimed at creating demand for sorghum products owing to the declining trend in the consumption of sorghum as staple food. Traditionally, consumption of sorghum is limited to *roti, kanji, mudde,* and *annam.* Preparation of sorghum-based products is cumbersome, time consuming, and requires efforts, as sorghum does not contain a protein called gluten (present in wheat), which makes the dough elastic and helps in rolling. However, this drawback has been removed by using the processing technologies (milling, flaking, baking, blending, and extrusion) developed by DSR, which helps in preparing sorghum-based recipes more conveniently. Sorghum products developed by DSR have lower glycemic index and glycemic load compared to wheat-based products and are nutritionally superior than refined wheat (*maida*) and rice-based products available in the market. These sorghum-based products are discussed in detail subsequently.

Jowar-rich multigrain flour

Jowar-rich multigrain flour is a blend of sorghum and other cereals such as wheat, *ragi*, pulses (black gram/soya), and fenugreek *(methi)*. The blended flour meets the required nutritional needs of the people who normally look for modern and healthy food. Different grains in this blend have varied advantages. Sorghum, other whole grains, and pulses add minerals, dietary fibre, and nutrients to *rotis*, making them nutritionally rich than the normal *rotis* made from wheat/*jowar*. Addition of wheat to the dough makes it pliable and allows flexibility in shaping while retaining the original flavour of wheat-based *roti*. Sorghum-rich multigrain flour can be stored for a period of two months at ambient temperature.

Jowar flakes

Sorghum flakes have been developed by DSR with the use of technology combining edge runner (flaking machine) and roaster. Sorghum flakes are RTE food, white in colour, and easily digestible. They contain higher protein, fibre, calcium, magnesium, iron, zinc, and folic acid as compared to rice flakes. They can be used for the preparation of variety of products such as *upma* or *poha,* porridge, *chuduwa*, fried and seasoned mixtures, and so on. Sorghum flakes can be stored for four months at ambient temperature.

Jowar rawa

Sorghum semolina, locally called *rawa* or *suji*, can be made through milling technology. DSR has developed three different types of *rawa* (coarse, medium, and fine) for the preparation of various types of food products. It is a RTC product with greater nutritional significance, especially in case of obese and diabetic patients. Sorghum *rawa* can be stored for a period of three months at ambient temperature. Availability of sorghum in the form of *suji* will help in overcoming drudgeries involved in preparing *suji*-based food.

Jowar biscuits

Biscuits are RTE food. DSR has prepared pure sorghum biscuits with sorghum flour, trans-free fat, sugar, natural flavouring agents, and baking powder. They are rich in protein, fibre, calcium, magnesium, iron, zinc, and folic acid and contain less sugar than the other biscuits available in the market. Pure sorghum biscuits can be stored for a period of six months.

Jowar vermicelli and pasta

Jowar vermicelli and pasta are RTC products, prepared through cold extrusion. The texture of these products becomes hard and brittle after drying, which can then be cooked in few minutes. *Jowar* vermicelli and pasta are prepared from sorghum and wheat semolina in the ratio of 7:3, respectively. Blending the product with wheat semolina offers visco-elasticity to the dough as sorghum lacks gluten. *Jowar* vermicelli and pasta improve the digestibility, control the sudden rise of blood glucose levels, and have nutritional benefits, thereby leading to their increased consumption. Sorghum vermicelli and pasta can be stored for six months at ambient temperature.

CHAPTER 5

Processing of Grain Sorghum: Technology Choices and Limitations

Food processing in all the sectors involves the conversion of raw material into a finished end-product of higher value to consumer. In some situations, processing is a one-step conversion of raw material to a consumer product. The history of food processing emphasises on establishing and maintaining microbial safety, as well as economic shelf-life of the food. All developments in food processing have similar origins. One common aspect is achieving and maintaining microbial safety in the product. Food processing methods are used worldwide as they improve nutritional quality of the grains, digestibility, and shelf-life. Food processing operations mainly involves primary and secondary processing steps.

Primary processing is a crucial step to maintain the quality of grain, which is a vital consideration in ultimately ensuring the quality of processed foods. In this process, stones, sand, dust, hay, glumes, animal excreta, and other foreign matter are removed. Grading and sorting are usually employed that aid in the removal of oversize, undersize, immature or poor quality grains. Secondary processing is a set of operations, wherein the processed raw material obtained after primary treatment is converted into ready-to-eat (RTE) and ready-to-cook (RTC) products. These finished products minimise the cooking time and are used as convenience foods.

Sorghum, the indigenous millet, is a traditional staple food in Africa, India, and other Asian countries. The consumption pattern of sorghum is quite traditional, and sorghum is not available as commercially processed food in the Indian market. This is due to lack of processing technologies, inavailability of machinery, and inconsistent grain availability. In semi-arid regions, dehulling and milling are done by hand pounding, and unfermented baked food and fermented food are used to prepare sorghum-based food. These methods are tedious, time-consuming, and manual, and almost entirely performed by women. The end products are of poor quality. Introduction of new technologies to process sorghum can remove the constraints involved in traditional way of sorghum processing while improving its nutrient content and shelf-life.

Storage Practices for Sorghum

Storage practices play a vital role in maintaining the quality of the product for further processing or consumption. Improper storage conditions/practices lead to high losses of raw material due to the attack of pests like weevils, beetles, moths, and rodents (Kartikeyan et al. 2009). According to the World Bank Report (1999), India's post-harvest losses accounted for 12–16 million metric tonnes of food grains each year. The monetary value of these losses amounted to more than Rs 50,000 crores per year (Singh 2010).

Traditionally, small quantities of sorghum are only stored in traditional bins or silos, but bulk storage is uncommon. More often, sorghum grains are stored in gunny bags that are placed on a wooden platform. Storage methods for sorghum involve the use of mud structures and modern bins usually made from straw of paddy/wheat, bamboo, mud, and bricks. Grains can be stored either indoors or underground (Channal, Nagnur, and Nanjayyanamath 2004).

Kharif (rainy season) sorghum is more susceptible to moulds and it has low storage life compared to post-rain harvested sorghum, which normally matures during the cold climate period and the grain quality of which is relatively superior, fetching higher price in the market. In India, *kharif* sorghum cultivation is characterised by hybrids while *rabi* sorghum is dominated by varieties. Hybrids are normally found to have low shelf-life. Coloured sorghum is more prone to insect damage than white sorghum due to its softer endosperm. Hence, *kharif* and coloured sorghum are especially taken care of while storing them. During storage, nutrient loss occurs due to the infestation of moulds and weevils, and enzymatic reactions. However, following proper drying and monitoring processes during storage, quality of grain can be retained with an enhanced shelf-life.

Sorghum Processing Technologies

Sorghum needs to be subjected to processing so as to obtain end products fit for human consumption. There is no established technological information on sorghum processing at pilot/industrial scale, like any other cereals. Most commonly used sorghum processing technologies are milling, popping, fermentation, and malting. Emerging technologies such as cold and hot extrusion, baking, and brewing, wet milling for starch separation also have applications in sorghum processing.

Processing methods have been initiated by the Directorate of Sorghum Research (DSR)-led consortium under the National Agricultural Innovation Project (NAIP). DSR has developed and fine-tuned different processing technologies such as extrusion, flaking, milling, parboiling, baking, popping, etc., to produce good quality of processed sorghum-based products for better utilisation of sorghum by making these products available year round as a convenient food, thus providing choice to the consumers. Similar interventions have been made by other R&D institutions and State Agriculture Universities (SAUs) to develop processing technology for sorghum and millets as a means of value addition.

Advantage of sorghum processing

Following are the advantages of processed sorghum-based food.

1. Value addition resulting in the utilisation of end-product.
2. Providing consumers a choice in the form of RTE/RTC products.
3. All time availability of processed products of sorghum.
4. Increase in shelf-life.
5. Increase in sorghum/millet consumption.
6. By creating consumer demand, there will be increased profits, leading to stabilisation of acreage under sorghum.

Dehulling

Dehulling/decortication/debranning is done primarily to remove the pericarp, or the outer layer. Normally, pestle and motor or hand pounding are used to dehull the grains. This method of processing is not efficient because part of the husk remains in the grain, and there is a considerable loss of grains. Improperly dehulled grains cannot be ground to fine quality flour.

The removal of fibrous layers from sorghum can be done mechanically. The prototype of dehuller consists of grinding discs and acts as an abrasive dehulling surface. The grain is dropped in the feeder from where it enters the dehulled chamber where the grinding stones decorticate the grain. The process and quality of dehulling can be checked through the inspection door. Thereafter, the grain is released from the discharging door, which is at the bottom of the machine. Pearling of sorghum grain removes bitterness, increases palatability when converted into flour (Desikachar 1982), and improves the quality of *roti* made out of the flour (Vimala et al. 1996).

Kernels with vitreous endosperm or with higher proportion of endosperm are more suitable for dehulling and yield higher fractions of dehulled grains, whereas soft endosperm does not give high yields, and the operational losses are more. But exposing the soft endosperm kernels to steam for few minutes may harden the endosperm and make it suitable for dehulling.

Some of the sorghum genotypes are characterised by high concentrations of phytochemicals, commonly termed tannin (Subramanian et al. 1983), which is not desirable during processing and end-product development. However, Indian sorghum genotypes are white in colour and are either free of tannin or contain a very insignificant quantity. During *kharif* season, rains at the time of harvest leads to the development of moulds and causes discoloration of the pericarp of the grain. Dehulling then becomes important to remove these moulds and phytochemicals from the grain so as to improve the quality.

Dehulling machines are available now to dehull the grains. However, nutrients are lost in the process as a result of the removal of pericarp (fibre, minerals), though coarseness of the grains is removed. It is thus recommended that partial dehulling be done so as to minimise the loss of nutrients. Dehulling after parboiling helps in the redistribution of micro-nutrients in the endosperm. Dehulling makes the end-products such as bakery (biscuits, cake, bread, etc.) fine in texture. The process followed during dehulling of sorghum is explained in Fig. 5.1.

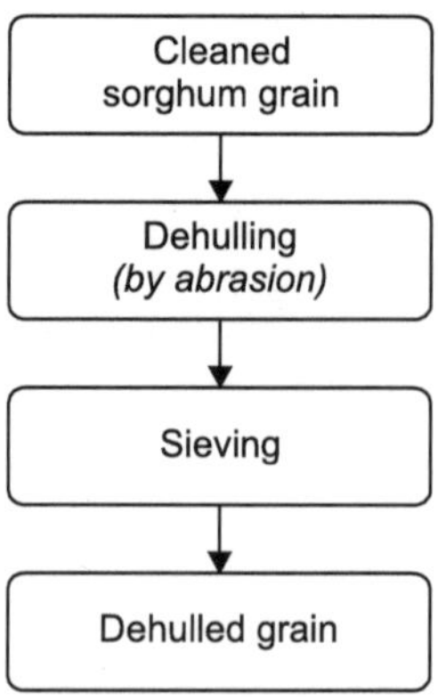

Fig. 5.1: Flow chart depicting dehulling of sorghum

Milling

Milling is a process of separating the bran and germ from the starchy endosperm so that the endosperm can be ground into flour and *rawa*. Milling in food grains can be done by wet or dry methods.

Wet milling in corn is usually done extensively to separate starch for food. This is done by steeping the sorghum grain in the aqueous solution with or without the antimicrobial agent sulphur dioxide (Shandera, Parkhurst and Jackson 1995). Dry milling involves cleaning and milling of the grain into the desired product. It includes the separation of the anatomical parts (bran and germ) of the grain and reducing the particle size of the endosperm.

Milling quality of sorghum can be determined by kernel shape and size, and characteristics and hardness of the endosperm. Sorghum milling can be done in a *chakki* (emery stones)/hammer/plate mill. For custom milling, emery or iron disc mills are used. At industrial level, hammer and roller millers are employed for this purpose. Milling yield depends on the type of cultivar, characteristics of endosperm, type of mill, and pre-treatment (3–5 per cent moistening of the grain before processing).

Flaking

RTE breakfast flakes are popular snacks made from cereals. Rice and corn flakes are produced and consumed across the world. Traditional process flaking involves pounding the paddy in a wooden mortar and pestle. *Poha,* or *chuduwa,* made from rice flakes is an important breakfast item in semi-urban and rural areas, and among the middle class families of urban India. There is also a fairly large market for flakes as it is extensively consumed all over the country round the year. However, sorghum flakes have been introduced very recently, which may trigger an overall demand for sorghum.

Flaking can be done for cereals and millets by conventional (edge runner), roller, flaker and, extrusion cooking methods. DSR has retrofitted the rice flaking machine for sorghum and optimised the conditions. The use of paddy flaking equipment, that is, edge runner (flaking machine) along with roaster has made it possible to produce flakes from sorghum very easily. In this method, sorghum grain is hydrated to equilibrium moisture content, incubated to remove the surface moisture, and subjected to high temperature short time (HTST) treatment. During HTST treatment, starch granules of the grain get gelatinised, and these grains are fed to edge runner, wherein the gelatinised endosperm gets flattened, and bran is eventually separated, pulverised, and collected from another outlet. Sorghum flakes obtained from this technology are similar to the rice flakes and are of better texture due to their quick hydration characteristics and less chewy properties. The yield recovery fractions from this technology depend on the cultivar type, characteristics of the endosperm, and the optimised conditions, and the resultant output is usually 50–65 per cent. The broken flake powder is a by-product, which can be used for making value-added sweets, which increases the overall profitability of flaking commercially.

Roller flaking is commonly used for making corn flakes, which is capital intensive as compared to the conventional edge runner. This technology involves pearling of grains and hydrating, their steaming to gelatinise the starch, and then flattening them between rollers. This type of flakes are different from the flakes obtained by conventional method in texture, as they are more leathery and chewy and more suitable for blistering or deep oil frying.

Flaking is done using extrusion technology, which produces breakfast cereals that can be consumed with cold or hot milk. This technology is widely used for corn. However, there have been some efforts to adopt this technology for sorghum flaking, which involves cooking, extruding, and shearing into small balls of the grain flour or *suji* followed by flattening in the roller flaker, drying, blistering, and finally flavour coating with or without sugar. The process explaining the production of sorghum flakes using edge runner is given in Fig. 5.2.

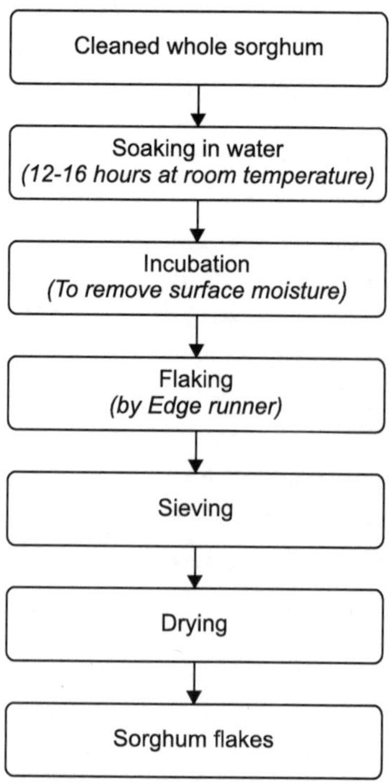

Fig. 5.2: Flow chart depicting sorghum flaking using edge runner

Baking

Baking technology is normally used to prepare a wide range of food products, including biscuits, cookies, breads, cakes, pastries, and crackers. The key ingredients in baked food are cereal flours such as wheat, maize, sometimes sorghum/millets in some quantity, sugar, fat, egg, and baking powder with food additives. In general, all flours are good sources of protein, energy, iron, and vitamins, but the nutrient content of the raw material and milling technology influence the nutrition in final products. As these ingredients undergo heat treatment, they experience changes in both form and structure.

Bakery products are becoming popular in developing countries as indicated by the fact that in the last decade the production of these products has more than doubled. The main reason for this may be fast changing food habits due to increasing urbanisation. Also, industrialisation coupled with easy availability, low cost, ready to eat characteristics, and high nutritional value of the products are other factors that account for the increased use of the bakery products (Kamaliya 2004).

Biscuits/cookies are some of the processed cereal products that have gained wide acceptance in India. These are ready to eat, convenient, and low-cost products containing nutrient of vital importance (Kulkarni 1997). Sorghum flour can also be used for bakery processing due to the lack of gluten protein in it (Rooney, Faubion and Earp 1980). Composite flours consisting of wheat and sorghum can be used in the preparation of such bakery products without affecting their texture and taste. Therefore, partially replacing wheat flour with sorghum flour is a possibility for increasing the utilisation of indigenous sorghum crops while contributing to the lowering of cost of bakery products (Ayo and Gaffa 2002). The underlining principle in biscuit making is to improve their value by improving their functional and health benefits, example being high-fibre biscuits.

Under NAIP, the DSR has standardised and optimised various types of biscuits made from pure sorghum and also from composite flour with different flavours. In this method, fat and sugar are blended in planetary mixer with a flat beater. Flour is mixed with baking powder and sieved for better mixing. And to the cream mixer, milk solids, flour mix, and essence are added, which are then well mixed to make dough. Dough is sheeted using a rolling pin and sheared with specific moulds to give a desirable shape followed by baking in an aluminium tray at 150°C for 25 minutes. After baking, biscuits are cooled and packed.

Sorghum extrusion

Extrusion of food is one of the popular technologies used by the food industries to make a large number of products of varying size, shape, texture, and taste. It is a process of forcing a material through a small restricted opening. It is essentially a thermo-mechanical process that involves mixing, kneading, extruding, cooling, forming, and partial drying or puffing operation steps. During extrusion, the raw materials are subjected to high temperature (50–250°C), pressure, and shear, which leads to the chemical and structural transformation such as starch gelatinisation, protein denaturation, inactivation of lipase enzyme, and degradation reactions of micro-nutrients like vitamins and pigments.

Food extrusion by definition is a process in which food material is forcibly pumped through small aperture, under various conditions of mixing, heating, and shearing, leading to the formation and/or puff drying of the ingredients. Extrusion is a highly versatile operation that can be applied to a variety of food processes.

There are several different types of extruders available in the market. A few examples include dry extruders, single screw extruders, interrupted-flight screw extruders, and twin-screw extruders. Single screw extruder consists of a live bin, feeding screw, pre-conditioning cylinder, extruder barrel, dye, and knife. In recent years, there is an increasing requirement for new and higher quality products for which single screw extruders are no longer adequate.

For these processing requirements, twin-screw technology must be used. Twin-screw extruders consist of several sub-components very similar to single-screw extruders but include various machines with widely different processing and mechanical characteristics and capabilities. Moisture content is critical during extrusion process for starch gelatinisation and protein denaturation. Twin-screw extruders have the ability to operate under the narrow or wide range of moisture.

Extruded products are RTC products, including vermicelli and pasta, and RTE products. Two types of extruders are in use: cold extruder and hot extruder. Hot extruder is more popular, used for making snacks (RTE) like Kurkure.

Extruded RTC products are also becoming popular in urban areas, which require less time for cooking. The products are made with sorghum flour or combination of sorghum flour and semolina, example being pasta and vermicelli.

Sorghum vermicelli

Vermicelli is a popular instant snack. These days, due to changing lifestyle and the influence of West, consumers prefer instant food. Because of this reason, vermicelli has good market and is preferred by people of all age groups. Under

NAIP, DSR has developed and standardised a process to manufacture vermicelli from sorghum in an effort to expand the product range and thereby increase the consumption of sorghum. This will ensure its popularity as a snack among all age groups, especially children. As sorghum is devoid of gluten, a protein that is responsible for imparting viscoelastic properties, wheat/gluten can be added to it to improve the texture and quality of end products.

Semolina of sorghum and wheat is mixed and hydrated with 25-30 per cent of potable water in a mixer, wherein blending and mixing takes place. This mixture is then subjected to extruder screw, wherein it withstands high temperature (55-65°C) and is pumped through a vermicelli dye. Adding of milk solids or vegetable pulp while preparing dough will enhance nutritive value of the end product.

Sorghum pasta

With the advance in technology, it has become feasible to produce commercial products from sorghum. Currently, only wheat pasta is available in the market. Sorghum pasta is prepared with sorghum semolina and wheat. Because of the lack of gluten content in sorghum, some amount of wheat needs to be added to prepare sorghum pasta.

The technology used for vermicelli can also be used for making pasta by changing the dye. Pasta made from sorghum can be made available in different shapes and sizes.

Gluten-free products can also be prepared using alternative binding agents, which would cater to celiac patients who are allergic to gluten protein. Fig. 5.3 gives an account of the preparation of sorghum pasta and vermicelli.

Extrusion cooking (or) hot extrusion

Extrusion cooking is a very popular, modern food processing technology used across the world for preparing snacks or supplementary food as well as pet food. In food processing, extrusion combined with the act of heating is used to create cooked products of varying shapes and sizes. Extrusion cooking can be done by twin-screw extruders. Commercially, most of the extruded snacks are prepared from corn, that is, sorghum grits, rice flour, and soya flour. The mixture passed through twin-screw extruder to produce expanded snacks at high temperature (> 100°C). The snacks can be coated with desired spices to provide varied tastes and flavours. They can also be extruded in a single extruder without puffing and can be dried and coated with a variety of spices and flavours (Fig. 5.4).

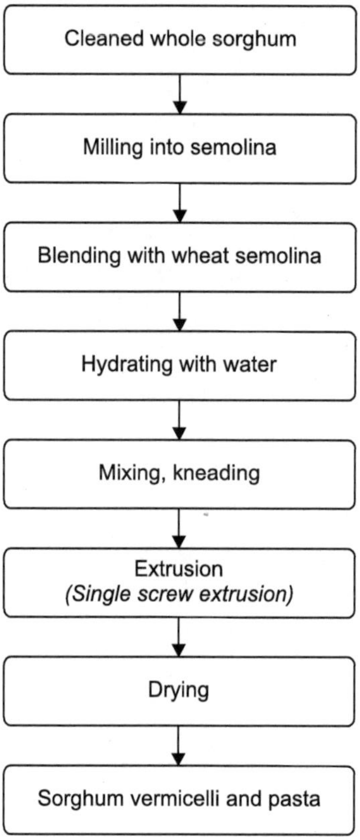

Fig. 5.3: Flow chart showing the preparation of sorghum pasta and vermicelli

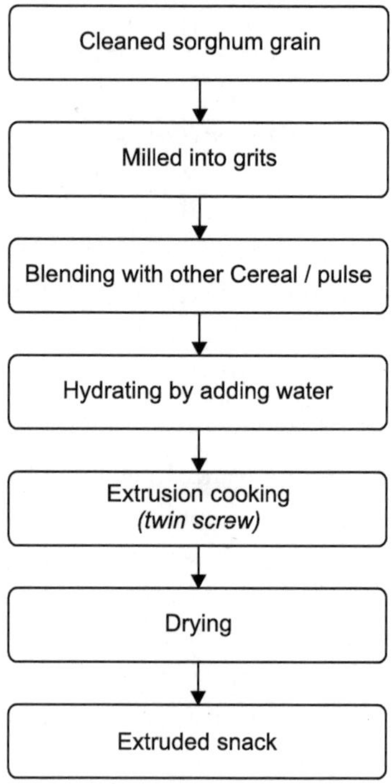

Fig. 5.4: Flow chart illustrating the preparation of extruded snack

Popping

Popping is simple, inexpensive, traditional processing technology, generally used for cereals and millets. It involves hydrating and incubating the grains followed by heat treatment. In this technology, high temperature allows grain starch to get pre-gelatinised in short time, and the pressure allows the endosperm to expand from bran. Popped grains develop highly desirable aroma and crunchiness. Also, lipase enzyme gets deactivated, which increases the shelf-life of the product.

Popping throughout the world generally involves HTST treatment, using conventional method of dry heat, sand and salt treatment, hot air popping, and microwave as heat transfer media (Yenagi et al. 2005). It is a popular snack among the people of all ages. Sand as heat transfer media is not an efficient method as sand usually adheres to the grain and thus affects the quality of the grain.

Generally popping can be done by equilibrating the sorghum grain to ~16% moisture and then exposing it to heat transfer media (240–270°C) for a few seconds. Introducing the grain to heat transfer media generates super heat vapour inside the grain leading to instantaneous heating, which cooks the grain and expands the endosperm, which then escapes with great force through the micro-pores of the grain. During this process, the grain gets sterilised and most of the seed microflora are destroyed. There is some kind of denaturation as well (Hadimani 1994).

Sorghum pops are consumed in several states of India as a snack. It has been found that sorghum pops are similar in flavour and nutrition to popcorn (Subramanian 1956). Small-sized grain with a dense and corneous endosperm is more suitable for sorghum popping (Ayyangar and Ayyer 1936). Popping improves starch and protein digestibility besides developing starch resistance and increasing albumin content in sorghum. Also, anti-nutritional inhibitors like phytic acid content get reduced during popping (Sandhya, Meera and Sila 2011).

Gun popping is also available in East Asian countries in grains with low amylose content. In this process, raw material withstands heated pressure in a closed container and then suddenly releases the product. The product expands and pops (Fig. 5.5).

Sorghum pops are part of traditional recipes and are popular in villages. However, efforts need to be made to popularise sorghum along the lines of popcorns.

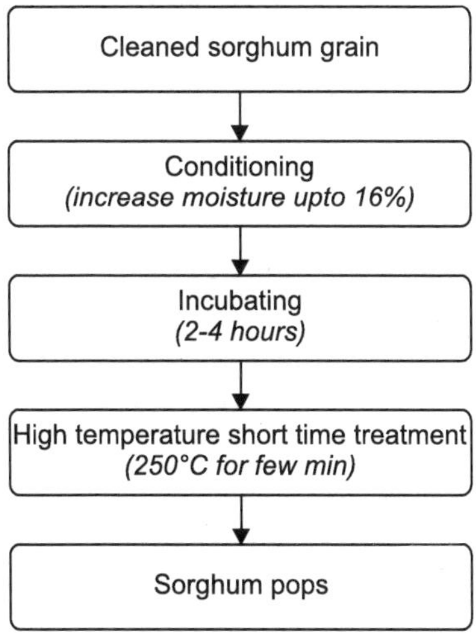

Fig. 5.5: Flow chart depicting the preparation of popped sorghum

Parboiling

Parboiling technology, commonly used for paddy, involves various steps, including soaking, steaming, and drying. This technology is adopted by various rice-growing countries, such as India, Bangladesh, Pakistan, Malaysia, Nepal, Sri Lanka, and South Africa.

Parboiling process generally hardens the endosperm, which subsequently increases the milling yield recoveries with reduce breakages/operational loss. The process involves soaking the grain in normal/hot water in order to saturate it with moisture. The soaked grain is then steam-heated till the starch granules get gelatinised. Then the dried grain is milled (Ayamdoo et al. 2013). Parboiling in paddy can reduce the nutrient loss during milling, especially by redistributing the minerals and heat-stable vitamin B complex from the outer layers to the inner layers of the grain (Sabry and Tannous 1961).

DSR has standardised and optimised the parboiling technology for sorghum. In this process, the grain is soaked in water to saturate it with the moisture, which is followed by steam cooking for 15 minutes. Then it is dried and used for further dehulling or milling. Using this technology, dehulling and milling yield fractions are increased and breakages are reduced. During dehulling, mineral loss will be more in sorghum due to the removal of pericarp, which can be overcome by parboiling before dehulling (Fig. 5.6).

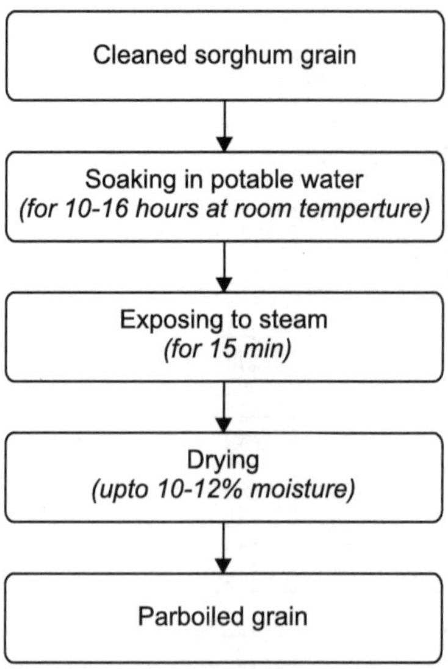

Fig. 5.6: Flow chart showing the manufacturing of parboiled *jowar*

Limitations

Inconsistent availability of sorghum grain in quantity and quality

Sorghum grain production and utilisation as food has declined in India due to the shift in cultivation from coarse to fine cereals and cash crops. Another important factor is government policy with respect to the fine cereals: rice and wheat. In India, rice and wheat are distributed through public distribution system (PDS), which has contributed to the decreased consumption of sorghum and millet.

Quality of sorghum grain is also essential for processing. But in India, grains of only *rabi* season are used for food because of their superior quality in terms of bold grains, white colour, and better taste. *Kharif* sorghum is inferior in quality, thus not good for food processing as it is prone to mould infestation and rust, and its quality gets deteriorated further in rains. Grain quality of seasonal sorghum determines the grain availability for industrial purposes.

Lack of knowledge of processing interventions

Processing technologies and machineries for sorghum and millets are not readily available. Lab models and prototype machineries have been developed by research and development institutes, agricultural universities, and colleges offering home science. It may be worth mentioning that small millets are marketed in large scale in Nasik, Maharashtra, and in Teni, Tamil Nadu. These mills are capital intensive and the yield is hardly 50 per cent. Hence, there is a need for the development of an integrated mill, which can meet the needs of private industries like Britannia and Parle.

Inadequate promotional awareness on health aspects

Sorghum and other millets are excellent sources of carbohydrates, proteins, micro-nutrients, and phytochemicals with nutraceutical properties. They contain 7–12 per cent proteins, 2–5 per cent fat, 65–75 per cent available carbohydrates, and 15–20 per cent dietary fibre. Higher level of dietary fibre, micro-nutrient content, complex nature of carbohydrates, and phytochemicals with health promoting properties ensure a place of pride for these coarse cereals as they are termed as 'healthy cereals'. Hence, the image of sorghum and millets as a poor man's food can be changed by developing products with improved nutrition and texture. It may be noted here that the urban consumers prefer convenient products with good taste, texture, colour, and longer shelf-life at affordable prices. In view of this, to make sorghum-based products more popular and easily available, innovative methods of sorghum processing need to be developed.

CHAPTER 6

Consumer Acceptability of Processed RTE/RTC Sorghum-based Products

About the Research

The consumer acceptability test is conducted under the study titled 'Sorghum: An Emerging Cash Crop: Consumer acceptability of processed RTC/RTE Health foods'. The core objective of the study was to understand the following parameters: consumer behaviour on the acceptance of the products, consumer profiling with delineation of specific target segments in selected cities, advising on positioning, providing inputs on branding, communication strategy, pricing, and distribution strategies. The study was conducted for five *jowar*-based food products, namely, *jowar*-rich multigrain *atta*, *jowar* vermicelli, *jowar rawa*, *jowar* flakes, and *jowar* biscuits.

Sampling and Methodology

The cities chosen for the study were Hyderabad, New Delhi, and Pune. Based on the statistical significance level, respondents/samples from over 400 household were covered in each city. Except *jowar* biscuits, all other products require cooking prior to consumption. Each household was provided with a sample kit containing *jowar*-rich multigrain *atta*, *jowar* vermicelli, *jowar rawa*, *jowar* flakes, and *jowar* biscuits.

Information on various factors was collected from the respondent in two stages, factors being packaging, product attributes before and after cooking, and experience while preparation and consumption. A structured questionnaire was used to get the information. In Stage-1, information pertaining to demographic profile (age, family size, dietary preference, income level, number of employed members, number of children, number of elders, and occupation of the lady members of the household) was collected. In Stage-2, the consumers were asked to provide their acceptability rating on the product attributes, including packaging, appearance, colour, smell, texture, preparation process, taste, etc. for both raw as well cooked products using 5-point hedonic scale (5 = liked very much, 4 = liked moderately, 3 = neither liked or nor disliked, 2 = disliked moderately, and 1 = disliked very much). Consumers rated the information on *jowar* and its benefits and the decision to purchase or not on binary (yes/no) scale.

The final outcome of the consumer response in each of the products is presented in this chapter, while the detailed analysis and other outputs of the analysis are included in the annexure.

Descriptions of Factors/Dimensions Used for the Analysis

- **Organoleptic:** This is the most important dimension identified by the respondents. It deals with the physical attributes and organoleptic characteristics of the product. Physical attributes include appearance of the product in raw form as well as during the preparation (cooking) and in the final product. Organoleptic attributes also include sensory aspects (cooked) like appearance, taste, aroma, texture, etc.
- **Prior awareness on health and other benefits of *jowar*:** This dimension includes awareness on benefits/perceptions attached to *jowar*. Parameters of this dimension include nutritional benefits, ease of digestion, source of energy, acceptance as modern food, etc.
- **Comparison of *jowar*-rich multigrain *atta* with traditional wheat *atta*:** The comparison of *jowar*-rich multigrain *atta* with traditional wheat on the basis of preparation, process difficulties, and test are the major attributes included under this dimension.
- **Socio-economic and value for money aspects:** Socio-economic and economic aspects of the respondents like income, expenditure on food, product pricing are covered under this dimension.
- **Demography:** Family size, number of children, family background, and occupations of the respondents are the attributes included in this dimension.
- **Myth related to *jowar*:** In many areas, there is a wrong perception that *jowar* is to be restricted only to the old, sick and adult members of the family. This myth on *jowar* consumption is also another dimension.

Cluster Analysis

Cluster analysis was carried out to classify the consumers based on their willingness to purchase the product. Following this, the clusters have been further grouped as 'Good', 'Average', and 'Bad' clusters for clarity in the output.

Each cluster is compared and classified based on the analysis of the consumer feedback using three parameters, namely, percentage of purchase vs non-purchase out of total, attribute likewise ranking, and perceptions on comparison with the traditional/current consuming products.

Jowar-rich Multigrain *Atta*

Table 6.1 lists the important dimensions of the consumer's response on *jowar*-rich multigrain *atta*

Table 6.1: Important factors/dimensions of the consumer's response on *jowar*-rich multigrain *atta*

Dimensions	*Cronbach's Alpha*	*Eigen Value*	*Variance Explained (%)*
Organoleptic	0.944	9.191	27.85
Prior awareness	0.926	5.169	15.66
Comparison with traditional products	0.768	2.143	6.5
Socio-economic and value for money	N.A.		
Demography	N.A.		

A common rule of thumb for dropping the least important factors from the analysis is the K1 rule. Factors with more than 1.0 as Eigen value were selected (Fig. 6.1). Based on this, demographic was the last factor with a value 1.340 to be selected. The reliability of the data and consistency in the responses received for Factor 1 was 0.944, which is much higher than the measuring point of 0.600. Similarly, Cronbach alpha for Factor 2 and Factor 3 was 0.926 and 0.768, respectively. The test could not be conducted for Factor 4 and Factor 5 as these consist of independent variables, and responses are in different scale. The per cent variance explained by these five factors is 62 per cent.

Organoleptic factor with Eigen value of 9.191 emerged as the major dimension of *jowar*-rich multigrain *atta*, as perceived by the consumers. In this, attributes of *chapati* or final product of *jowar*-rich multigrain *atta* is more important as compared to the preparation and appearance parameters in the raw form. Prior awareness on the *jowar* products and their nutritional and health benefits are second important dimensions, followed by comparison with traditional wheat *atta*.

The factors mentioned above were grouped and named on the basis of the characteristics of the composite variables. The grouping termed as factor on the characters is discussed subsequently.

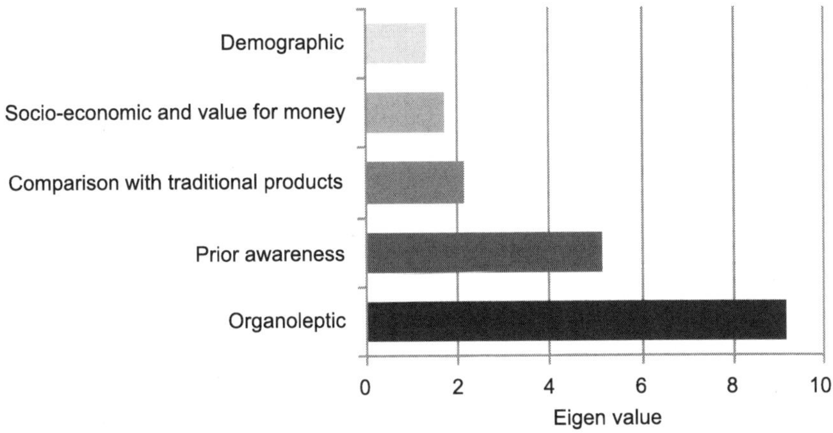

Fig. 6.1: Factors for consideration in *jowar*-rich multigrain *atta*

Organoleptic evaluation of jowar products

This is the most important dimension identified by the respondents. This deals with the physical attributes and organoleptic characteristics of the product. Physical attributes include product appearance in the raw form as well as during the preparation (cooking) and in finished form. Organoleptic attributes also include sensory aspects (cooked) like appearance, taste, aroma, texture, etc.

Prior awareness on health and other benefit of jowar

This dimension includes awareness on the consumption of *jowar* and benefits and perceptions attached to it. Awareness on nutritional benefits, ease in digestion, a good source of energy, a modern food, etc. are the parameters grouped in this dimension.

Comparison of jowar-rich multigrain atta with traditional wheat atta

The comparison of *jowar*-rich multigrain *atta* with traditional wheat on the basis of the difficulties encountered during the preparation process and taste is the major attribute included under this dimension.

Socio-economic and value for money

Economic aspects of the respondents like income, expenditure on food, and product pricing are the important aspects that are covered under this dimension.

Demography

Family size, number of children, family background, and occupations of the respondents are the attributes included in this dimension.

Similar dimensions have been found in the factor analysis for remaining products like *jowar* biscuits, vermicelli, *rawa*, and flakes. To avoid repetition, a common reference text has been placed before the result discussion in chapter 6.

Cluster of the respondents

Further, a cluster analysis was carried to classify the consumers into segments based on their willingness to purchase the product. Following this, the clusters were further grouped as 'Good', 'Average', and 'Bad' clusters for clarity in the output. Table 6.2 and Figs 6.2-6.4 give a representation of the major characteristics and features of the 'Good' and 'Average' clusters.

Table 6.2: Consumer response on purchase decision (*atta*), by cluster

Cluster	*Purchase*	*Not Purchase*	*Total Respondents*	*Cluster Output*
1	206	29	235	Average
2	6	23	29	Bad
3	24	21	45	Bad
4	4	24	28	Bad
5	817	19	836	Good
6	41	0	41	Good

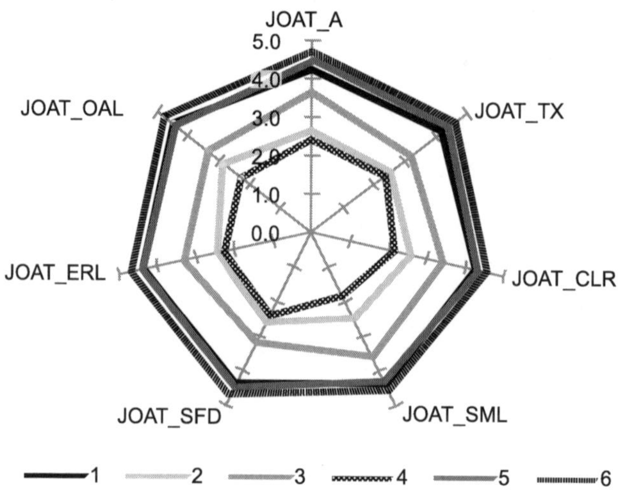

(Scale: 5 = Liked very much, 1 = not liked at all)

Fig. 6.2: Cluster profile based on product attributes (*atta*)

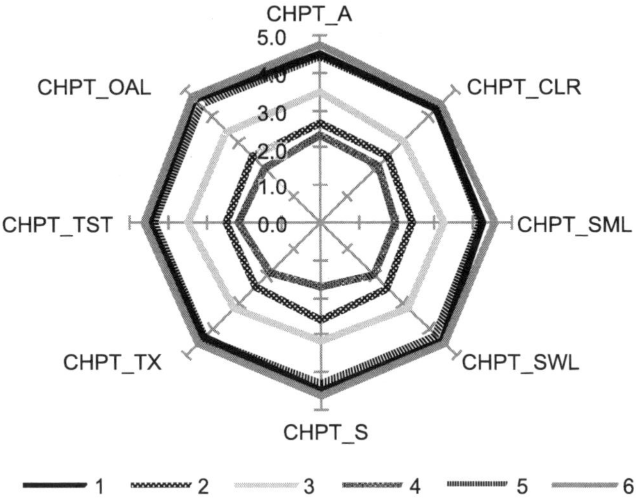

(Scale: 5 = Liked very much, 1 = not liked at all)

Fig. 6.3: Cluster profile based on product attributes *(chapati)*

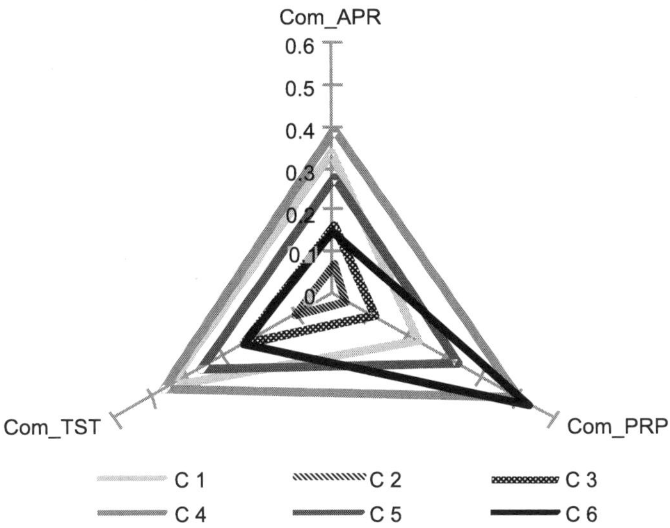

(Score: 0 = Traditional product is better, 1 = Jowar is better or both are same)

Fig. 6.4: Cluster profile based on comparison with wheat *atta*

The important characters of both Good and Average clusters are presented subsequently. The output of the cluster is further statistically tested (Chi-square and Probability [*P* value]) to know the significant similarities and differences between both clusters.

The result of Chi-square test and probability test on the socio-economic parameter on the clusters shows that there are significant differences (< 0.05) between both the clusters in terms of family size, children, food habit, household income, and willingness to pay the price. On the other hand, the clusters do not have significant differences in terms of occupation of the female respondents and household expenses on food from their total income. Detailed table of consumer profile from both good and average cluster is given in the Annexure.

Similar technique was followed for analysing the clusters for the rest of the products, as discussed later in this chapter.

Attributes influencing the purchase of jowar-rich multigrain atta

In the factor analysis, we found out important dimensions from the response of the consumers with no dependent variable (purchase/not purchase). The ultimate aim was to know the important attributes, which may or may not have an impact on the purchasing decision of the consumer.

A Logistic Regression Analysis (LRA) model was developed and stepwise analysis was done using various statistical tools (Table 6.3).

Table 6.3: Variables/attributes influencing the purchasing decision

Factor	*MLE Estimate*	*Odds Ratio Estimate*
Intercept	-9.2045	NA
Stat_1	2.4959	12.133
JOAT_ERL	0.4605	1.585
CHPT_CLR	0.3395	1.404
CHPT_SWL	0.4801	1.616
CHPT_TX	0.6327	1.883
CHPT_TST	0.6285	1.875

Note: The analysis of maximum likelihood estimates (MLE) was used to obtain parameter estimates. Significance of parameter estimates was based on the Wald Chi-square value at $P < 0.05$.

The Odds Ratio Score (ORS) estimates indicate the probability that by how many times the decision to purchase would increase if there is 1 unit increase in liking scale (Likert scale scores 1 through 5). Statement 1 (*jowar* provides wholesome nutrition to the entire family) emerged as the most important attribute, which could increase the chances of purchase. The response collected was on the non-numeric scale (agree/disagree) with the interpretation that increasing awareness on the benefits of *jowar* would certainly increase the chances of its purchase by the consumers. This could in turn be the interpretation of the ORS. On the other hand, *chapati* texture and taste are other most influential variables that could increase the purchasing chance by almost 1.8 times. Swelling of *chapatis* and the ease of rolling the *chapati* could increase the purchasing chance by 1.61 times and 1.58 times, respectively, when there is an increase in 1 unit in the liking scale.

Prediction of correct purchase decision

Table 6.4: Correct classification of actual purchase decision and prediction on correctness of purchase for *jowar*-rich multigrain *atta*

	No. of Respondents			*% of Respondents*		
	Not Purchase	*Purchase*	*Total*	*Not Purchase*	*Purchase*	*Total*
Not Purchase	59	57	116	5%	5%	10%
Purchase	21	1078	1098	2%	88%	90%
Total	80	1135	1214	7%	93%	100%
Correct Prediction			1137	Correct Prediction		93%
Incorrect Prediction			78	Incorrect Prediction		7%

Table 6.4 indicates that of the 1137 respondents, 1078, or 88 per cent of the total consumer respondents, are willing to purchase and 59, or 5 per cent of the consumer respondents who are not interested in purchasing the *jowar*-rich multigrain *atta* are observed and predicted as correct classification into a group. On the other hand, 57, or 5 per cent of the total consumer respondents who are willing to purchase and 21, or 2 per cent of the consumer respondents who are not interested in purchasing the *Jowar*-rich *multigrain atta* are observed and predicted as incorrect classification group. In conclusion, the purchasing decision of *jowar*-rich multigrain *atta* could be predicted with 93 per cent of accuracy and 7 per cent of inaccuracy.

Purchase prediction accuracy based on individual attributes (hit ratio)

Based on the LRA at *P* event of 0.05, Hit rate (%) is the correct classification of an unknown unit (product) into a group (purchased vs not purchased).

Chapati texture alone can be used for predicting the decision to purchase *jowar*-rich multigrain *atta* with the accuracy of 76.5 per cent. Good swelling of *chapati* (roasting on direct fire or after putting oil drops) is another important attribute/factor, which can alone be used as a single predictor, and purchasing prediction can be done with 73.3 per cent accuracy. Ease in rolling of *chapati*

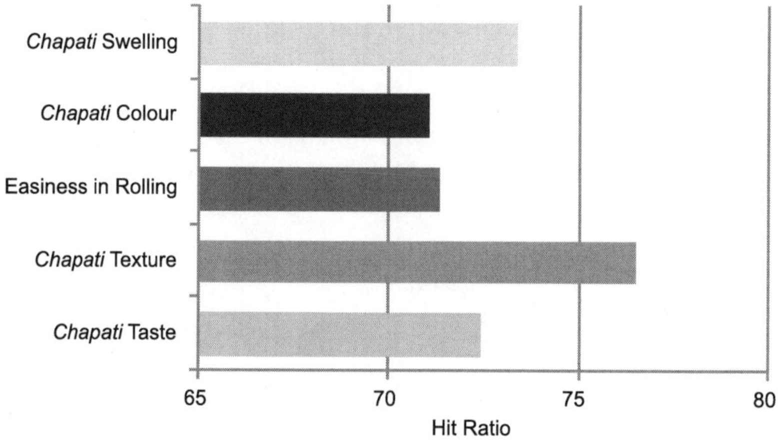

Fig. 6.5: Purchase prediction accuracy based on individual attributes for *jowar*-rich multigrain *atta*

and colour can provide accuracy of predicting the correct purchase decision by 71.3 and 71 per cent, respectively (Fig. 6.5).

Similarities and dissimilarities in the response of the respondents on important attributes of jowar-rich multigrain atta

Surveyed cities were grouped in pairs and analysed for attributes influencing purchase (Fig. 6.6). The objective was to find out which pair of the city was similar or different on the level of response. Paired comparison test was

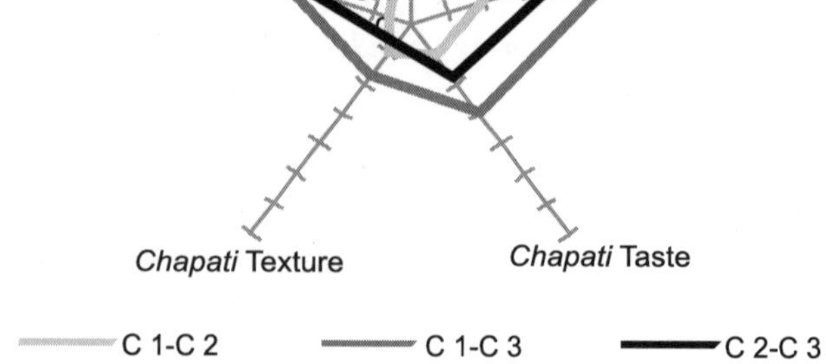

(Score: 0 = Traditional product is better, 1 = Jowar is better or both are same)

Fig. 6.6: Pair wise response of cities on influential attributes of *atta*

conducted using ANOVA techniques. Results of this analysis help in deciding the communication strategy to market *jowar*-rich multigrain *atta* in a particular city (* 95per cent confidence level [< 0.05]).

Hyderabad is symbolised as City 1, Delhi as City 2, and Pune as City 3. Hyderabad and Delhi do not have significant difference on the response received for the taste of *chapati*, *chapati* texture, and ease of rolling *chapatis*. Hyderabad and Delhi and Hyderabad and Pune are very close in their response on the *chapati* texture, but when comparing with the former pair of cities, the difference in response is little wide. Delhi and Pune appeared as different pairs having significant differences among all the variables as compared to former two pairs of cities.

Respondents of Hyderabad liked the overall attributes of the *jowar*-rich multigrain *atta* compared to the respondents of Pune and Delhi. When considering the most influential attributes of *jowar*-rich multigrain *atta*, Hyderabad respondents are near to extremely liking in all the attributes. In Delhi, respondents are satisfied with the swelling, texture, and taste of *chapati*, but colour and ease of rolling the *chapati* have performed average on the liking scale. In Pune, colour, taste, texture, and swelling of *chapati* have been liked moderately by the respondents, though ease of rolling *chapati* is liked in above moderate level on the liking scale (Fig. 6.7).

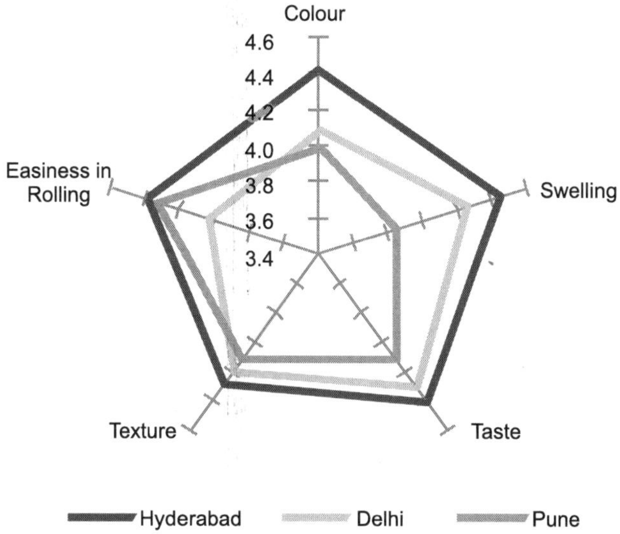

(Scale: 5 = Liked very much, 1 = Not liked at all)

Fig. 6.7: Impact of important attributes in response of individual city

Jowar Biscuits

Important factors/dimensions of the consumer's response on *jowar* biscuits are given in Table 6.5.

Table 6.5: Important facors/dimensions of the consumer's response on *jowar* biscuits

Dimensions	*Cronbach's Alpha*	*Eigen Value*	*Variance Explained (%)*
Organoleptic	0.896	5.91	21
Prior awareness	0.913	4.08	15
Comparison with traditional products	0.748	1.99	7
Socio-economic	N.A.	1.66	6
Myth of *jowar*	N.A.	1.43	5
Demography	N.A.	1.27	4.5

Demographic factor having the Eigen value of 1.27 was the last variable that had been selected. The reliability of the data and consistency in the responses received for Factor 1 are equivalent to 0.896, which is much higher than the measuring point of 0.600; however, the value is higher for Factor 2, which can be explained as a more reliable response. The per cent variance explained by these six factors is 58.5 per cent.

Organoleptic factor is an important dimension for the consumers to like or dislike the biscuits followed by prior awareness on health and other benefits of *jowar*. Comparison with other cookies that they consume is the third important factor in evaluating the *jowar* biscuits. Ultimately, the final experience of the consumer in terms of sensory and comparative attributes is going to be the major factor for accepting or rejecting the biscuit (Fig. 6.8).

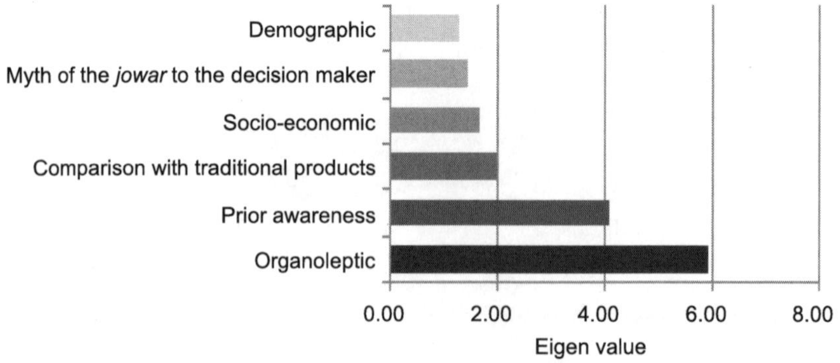

Fig. 6.8: Factors for consideration in *jowar* biscuits

Clusters of the respondents

The major characteristics and features of the Good and Average clusters are presented in Table 6.6 and Figs 6.9 and 6.10.

Table 6.6: Clusters of the consumers based on response on purchase decision

Cluster	*Purchase*	*Not Purchase*	*Total Respondents*	*% Purchase*	*Cluster Output*
1	6	40	46	13	Bad
2	43	0	43	100	Good
3	153	11	164	93	Average
4	199	4	203	98	Good
5	731	28	759	96	Good

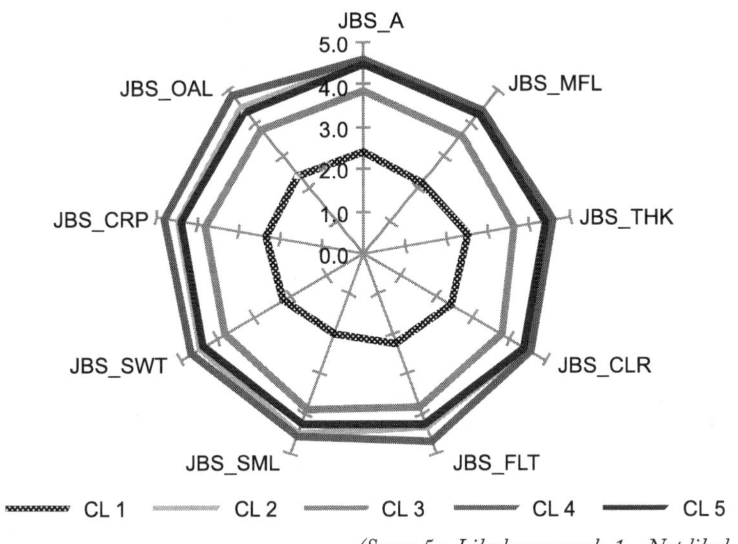

(Score: 5 = Liked very much, 1 = Not liked at all)

Fig. 6.9: Score on biscuits attributes, by cluster

The total respondents were formed into five clusters. Cluster 5 had only four respondents. Each cluster was compared and classified based on the analysis of the consumer feedback through three parameters, namely, percentage of purchase vs non-purchase, attribute likewise ranking, and perceptions on comparison with the traditional/current consuming biscuits.

The biscuit cluster widely differs from the cluster of multigrain *atta*. The clusters are significantly different in terms of price, buying preference, quantity

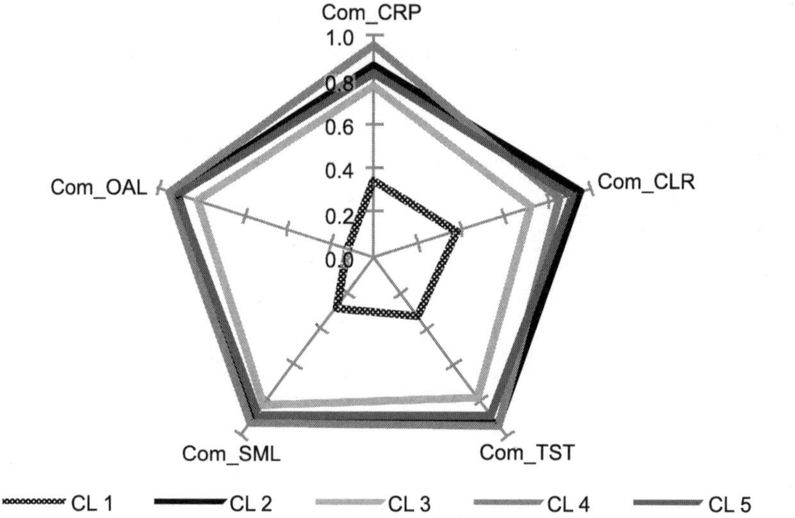

(Score: 0 = Traditional biscuit is better, 1 = Jowar is better or both are same)

Fig. 6.10: Score on comparison of biscuit attributes with traditional biscuits, by cluster

of biscuit consumed per month, percentage expenditure on food from the total household income, monthly household income, food habits, and number of children. Table detailing the consumer profile from both Good and Average clusters is given in the Annexure.

Attributes influencing the purchase of jowar biscuits

Table 6.7: Odds Ratio Estimate to predict purchase influencing attributes of *jowar* biscuits

Factor	*MLE Estimate*	*Odds Ratio Estimate*
Intercept	-6.2408	NA
JOBS_SWT	0.9833	2.673
JOBS_MFL	0.5173	1.677
Com_SML	1.4114	4.102
JOBS_CRS	0.5185	1.68

Note: The analysis of maximum likelihood estimates (MLE) was used to obtain parameter estimates. Significance of parameter estimates was based on the Wald Chi-square value at $P < 0.05$.

The smell of the *jowar* biscuit in comparison with the traditional biscuit is an important factor/attribute followed by its sweetness that could influence the purchasing decision of the consumers. As shown in Table 6.7, these two attributes, namely smell and sweetness, could increase the purchasing chance of *jowar* biscuit by 4.10 and 2.67 times, respectively, when there is an increase in 1 unit in liking scale. Taste of the biscuit in mouth is another important attribute that could increase the purchasing chance by 1.67 times followed by crispiness that increases the purchasing chance by 1.68. Awareness related to the benefits of *jowar* does not have a significant influence on the purchasing consideration.

Prediction of correct purchase decision

To draw an inference, 1114, or 92 per cent of the total respondents who are willing to purchase and 37, or 3 per cent of the respondents who are not interested in purchasing the *jowar* biscuits are observed and predicted as correct classification into a group. On the other hand, 46, or 4 per cent of the total respondents who are willing to purchase and 18, or 1 per cent of the respondents who are not interested in purchasing the *jowar* biscuit are observed and predicted as incorrect classification group. In conclusion, the purchasing decision of *jowar* biscuit could be predicted with 95 per cent accuracy and 5 per cent inaccuracy (Table 6.8).

Table 6.8: Correct classification of actual purchase decision and prediction on the correctness of purchase for *jowar* biscuits

	No. of Respondents			*% of Respondents*		
	Not Purchase	*Purchase*	*Total*	*Not Purchase*	*Purchase*	*Total*
Not Purchase	37	46	83	3%	4%	7%
Purchase	18	1114	1132	1%	92%	93%
Total	55	1160	1215	5%	95%	100%
Correct Prediction			1151	Correct Prediction		95%
Incorrect Prediction			64	Incorrect Prediction		5%

Purchase prediction accuracy based on individual attributes (hit ratio)

Sweetness of the biscuit can be used as the single predictor, and decision of purchasing *jowar* biscuit can be predicted with an accuracy of 71.5 per cent. Similarly, taste in mouth and crispiness of the biscuits can also be used for predicting the purchase decision by 65.8 per cent and 64.9 per cent accuracy, respectively (Fig. 6.11).

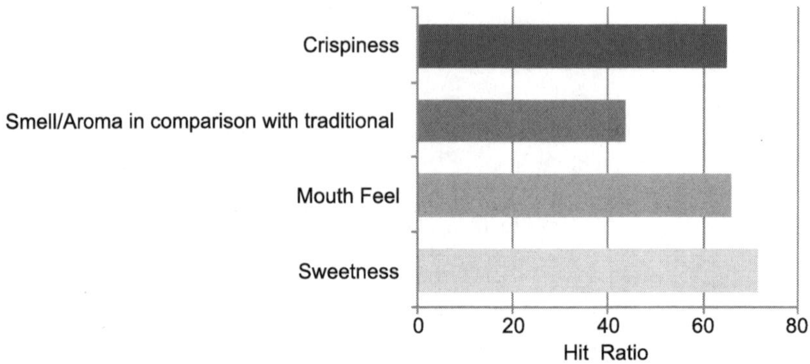

Fig. 6.11: Purchase prediction accuracy based on individual attributes for *jowar* biscuits

Similarities and dissimilarities in the response of the respondents on important attributes of jowar biscuits

Hyderabad is symbolised as City 1, Delhi as City 2, and Pune as City 3. Hyderabad and Delhi do not have significant difference in terms of the response on crispiness and sweetness attributes of *jowar* biscuit. Similarly, responses from Delhi and Pune are similar on taste in the mouth attribute. All cities have significant difference in the response in terms of smell/aroma attribute when compared with traditional cookies (Fig. 6.12).

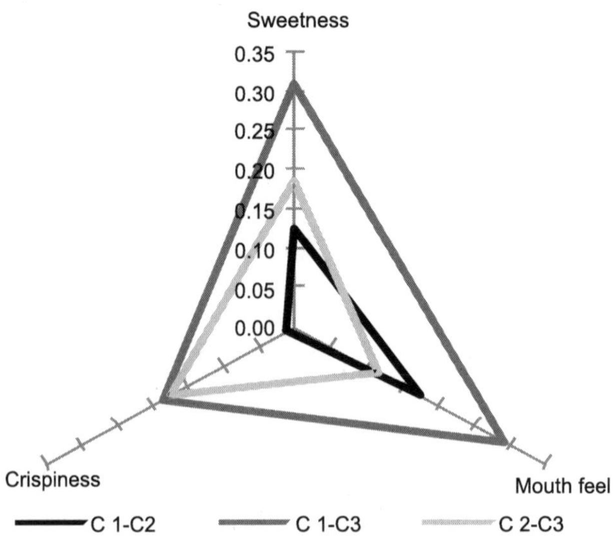

(Scale: 0 = Traditional product is better, 1 = Jowar is better or both are same)

Fig. 6.12: Pair wise response of cities on influential attributes of *jowar* biscuits

Hyderabad respondents like all the attributes of the *jowar* biscuit as compared to the respondents of Delhi and Pune. Taste in mouth and sweetness of the *jowar biscuits* are not liked much by the Pune respondents. Delhi and Hyderabad respondents have liked above moderately the crispiness and sweetness of *jowar biscuits*. On the other hand, respondents of Pune are nearly satisfied with identified attributes of *jowar* biscuits (Table 6.9; Fig. 6.13).

Table 6.9: Responses on influential attributes of *jowar* biscuit, by city

	Hyderabad		*Delhi*		*Pune*		*Chi-square*	*Probability*
	0	*1*	*0*	*1*	*0*	*1*		
Smell/aroma in comparison with traditional product	30	405	74	301	72	333	32.10	< 0.0001

(*Scale*: 1 = Jowar product is better or both are same, 0 = Traditional product is better)

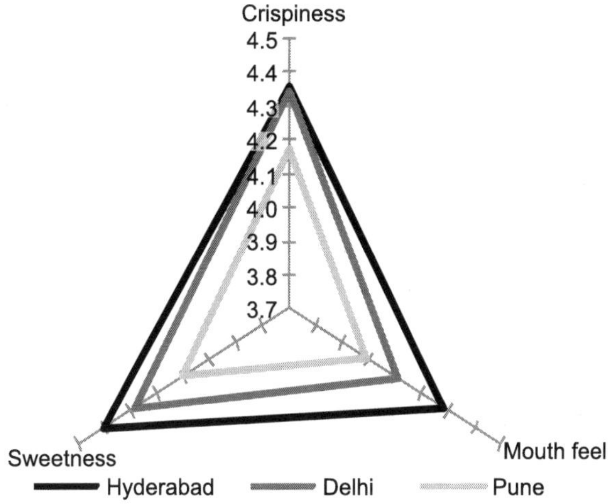

(*Scale*: *5 = Liked very much, 1 = not liked at all*)

Fig. 6.13: Impact of important attributes in response of individual city

Jowar Vermicelli

The reliability of the data and consistency in the responses received for Factor 1 is 0.917, which is much higher than the measuring point of 0.600. Cronbach alpha for Factors 1, 2, and 3 is 0.950, 0.914, and 0.699 respectively. The per cent variance explained by these six factors is 59.5 per cent (Table 6.10; Fig. 6.14).

Table 6.10: Important factors/dimensions of the consumer's response on *jowar* vermicelli

Dimensions	*Cronbach's Alpha*	*Eigen Value*	*Variance Explained (%)*
Organoleptic	0.950	8.82	28
Prior awareness	0.914	4.02	13
Comparison with traditional products	0.699	1.81	6
Socio-economic and value for money	N.A.	1.65	5
Myth of *jowar* to the decision maker and income	N.A.	1.36	4.5
Demographic	N.A.	1.28	4.0

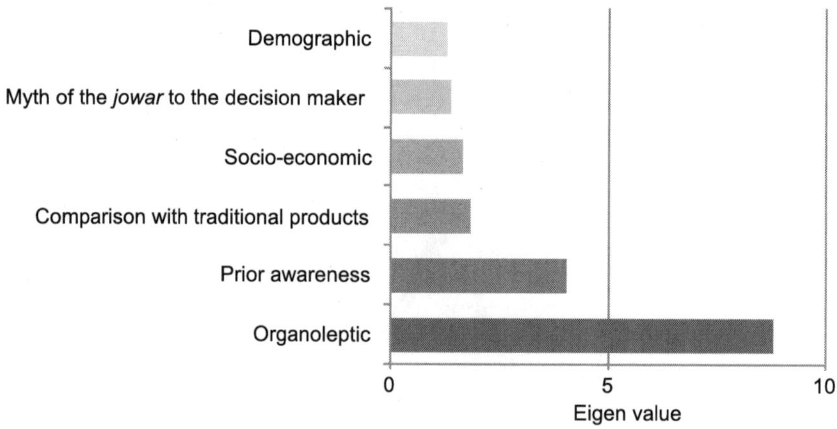

Fig. 6.14: Factors for consideration in *jowar* vermicelli

Clusters of the respondents

The *jowar* vermicelli clusters are significantly different in terms of place of majority purchase, buying preferences (branded or unbranded), occupation of the male member of the family, and size of the family (total members in a household) (Table 6.11; Figs 6.15-6.17). The table detailing the consumer profile from both Good and Average clusters is given in the Annexure.

Table 6.11: Clusters of the consumers based on the response on purchase decision

Cluster	*Purchase*	*Not Purchase*	*Total Respondents*	*% Purchase*	*Cluster Output*
1	555	29	584	95	Good
2	7	50	57	12	Bad
3	166	29	195	85	Average
4	325	51	376	86	Average

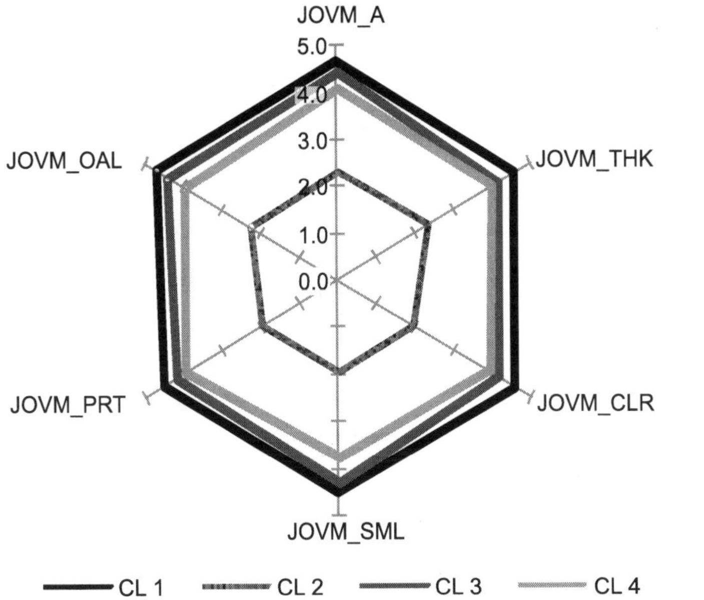

(Score: 5 = Liked very much, 1 = Not liked at all)

Fig. 6.15: Score on raw vermicelli's attributes, by cluster

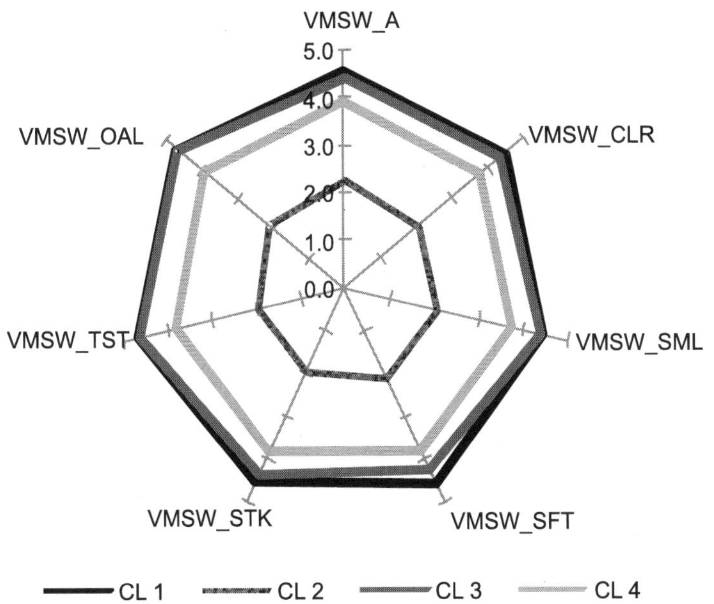

(Score: 5 = Liked very much, 1 = Not liked at all)

Fig. 6.16: Score on cooked (sweet) vermicelli's attributes, by cluster

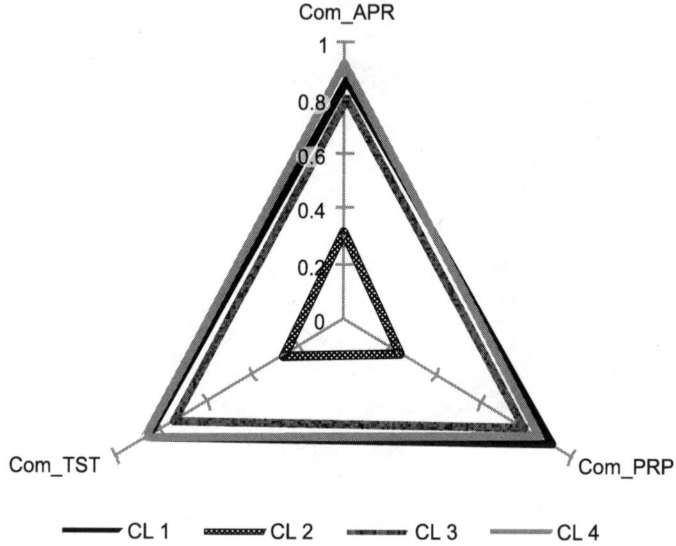

(Score: 0 = Traditional product is better, 1 = Both are same)

Fig. 6.17: Score on attributes of vermicelli in comparison with traditional vermicelli, by cluster

Attributes influencing the purchase of jowar vermicelli

The analysis of Maximum Likelihood Estimates (MLE) was used to obtain parameter estimates. Significance of parameter estimates was based on the Wald Chi-square value at $P < 0.05$.

Taste of *jowar* vermicelli in comparison with wheat vermicelli is the most influential attribute affecting the purchase decision. This attribute alone can increase the chances of the purchase of *jowar* vermicelli by 6.26 times when there is a 1 unit increase in the liking scale. Similarly, ease of preparation similar to traditional vermicelli could increase the purchasing chance by 3.24 times and the overall appearance of the *jowar* vermicelli just like traditional vermicelli could influence the purchasing decision by 2.45 times if there is 1 unit increase in the liking scale or if the appearance is similar to that of wheat vermicelli. Smell/aroma and stickiness of the cooked vermicelli could also enhance the purchasing chance by 2.26 and 1.49 times, respectively. Table 6.12 gives the Odds Ratio Estimate to predict the attributes influencing the purchase of *jowar* vermicelli.

Table 6.12: Odds Ratio Estimate to predict the attributes influencing the purchase of *jowar* vermicelli

Factor	*MLE Estimate*	*Odds Ratio Estimate*
Intercept	0.8243	NA
Smell of *jowar* vermicelli	0.8181	2.266
Stickiness in sweet vermicelli	0.3982	1.489
Sweet vermicelli overall	0.5142	1.672
Com_APR	0.8984	2.456
Com_PRP	1.1763	3.242
Com_TST	1.8347	6.263

Prediction of correct purchase decision

A total of 1028, or 85 per cent of the total consumer respondents who were willing to purchase and 84, or 7 per cent of the consumer respondents who were not interested in purchasing the *jowar* vermicelli were observed and predicted as correct classification into a group. On the other hand, 75, or 6 per cent of the total consumer respondents who are willing to purchase and 28, or 2 per cent of the consumer respondents who are not interested in purchasing the *jowar* vermicelli were observed and predicted as incorrect classification group. In conclusion, the purchasing decision of the *jowar* vermicelli could be predicted with 92 per cent of the accuracy and 8 per cent of inaccuracy (Table 6.13).

Table 6.13: Correct classification of actual purchase decision and prediction on correctness of purchase for *jowar* vermicelli

	No. of Respondents			*% of Respondents*		
	Not Purchase	*Purchase*	*Total*	*Not Purchase*	*Purchase*	*Total*
Not Purchase	84	75	159	7%	6%	13%
Purchase	28	1028	1056	2%	85%	87%
Total	112	1103	1215	9%	91%	100%
Correct Prediction			1112	Correct Prediction		92%
Incorrect Prediction			103	Incorrect Prediction		8%

Purchase prediction accuracy based on individual attributes (hit ratio)

Fig. 6.18 gives the purchase prediction accuracy based on individual attributes for *jowar* vermicelli.

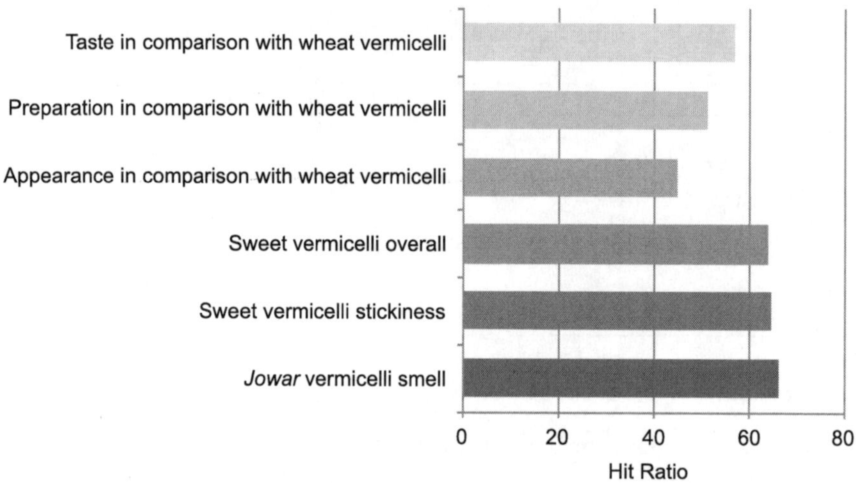

Fig. 6.18: Purchase prediction accuracy based on individual attributes for *jowar* vermicelli

Similarities and dissimilarities in the response of the respondents on important attributes of jowar vermicelli

Hyderabad is symbolised as City 1, Delhi as City 2, and Pune as City 3. Delhi and Pune have similar response on the stickiness attribute of vermicelli. In terms of the overall attributes, Hyderabad and Delhi have similar response. Hyderabad-Delhi and Delhi-Pune have similar response on smell/aroma of *jowar* vermicelli. The response on the attributes of *jowar* vermicelli when comparing with wheat vermicelli is also significantly different among the three cities, though majority of the respondents were of the opinion that *jowar* vermicelli is better than or not much different from wheat vermicelli (Table 6.14; Fig. 6.19).

Table 6.14: Response on influential attributes of *jowar* vermicelli, by city

	Hyderabad		*Delhi*		*Pune*		*Chi-square Test*	*Probability*
	0	*1*	*0*	*1*	*0*	*1*		
Taste in comparison with wheat vermicelli	60	375	113	262	52	353	48.62	< .0001
Appearance in comparison with wheat vermicelli	48	387	99	276	37	368	54.06	< .0001
Preparation in comparison with wheat vermicelli	48	387	95	280	51	354	35.85	< .0001

(*Score: 1 = Jowar product is better or both are same, 0 = Traditional product is better*)

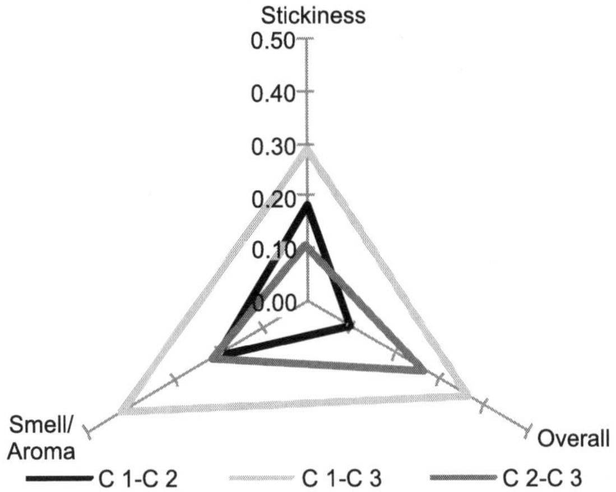

(Score: 0 = Traditional product is better, 1 = Jowar is better or both are same)

Fig. 6.19: Pair wise response of cities on influential attributes of *jowar* vermicelli

Pune respondents did not like the smell of *jowar* during the preparation and consumption of sweet vermicelli. Hyderabad again emerged as the city that liked all the attributes of *jowar* vermicelli. On the liking scale, liking of the vermicelli among the respondents of Hyderabad and Delhi is above moderate level. Pune again remains conservative as the score on the liking scale of the vermicelli attributes was an average (Fig. 6.20).

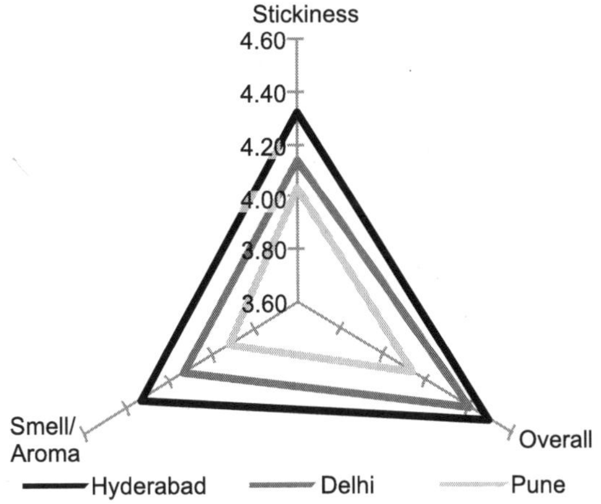

(Scale: 5 = Liked very much, 1 = Not liked at all)

Fig. 6.20: Analysis of the response of members of individual city with respect to important attributes

Jowar Suji/*Rawa*/Semolina

Suji/*rawa*/semolina is used for making different types of dishes in Pune, Delhi, and Hyderabad. In Hyderabad, respondents make *upma*/*uppit* with *rawa*, whereas Delhi makes *sira*/*halwa*/sweet. In Pune, respondents prepare both *sira* and *upma*. Both types of dishes have different attributes in terms of preparation and taste. To map the attributes of the respondents correctly, the product was divided into two groups and analysed.

Uppit/ *Upma*

The reliability of the data and consistency in the responses received for Factor 1 pertain to 0.946, which is much higher than the measuring point of 0.600; the value is lower for Factors 2 and 3, which can be explained comparatively as a less important response. The per cent variance explained by these six factors is 56.25 per cent (Table 6.15).

Table 6.15: Important factors/dimensions of the consumer's response on *suji*/*rawa*/semolina

Dimensions	*Cronbach's Alpha*	*Eigen Value*	*Variance Explained (%)*
Organoleptic	0.946	9.38	29
Prior awareness	0.720	2.58	8
Comparison with traditional products	0.697	2.45	7.5
Socio-economic and value for money	N.A.	1.57	5
Myth associated with *jowar*	N.A.	1.50	4.75
Demography	N.A.	1.40	4.0

Myth related to *jowar* that it should be restricted to unwell and old members of the family is the new dimension, which appeared to be an important factor for the consumer in deciding whether to consume *jowar* in their household (Fig. 6.21).

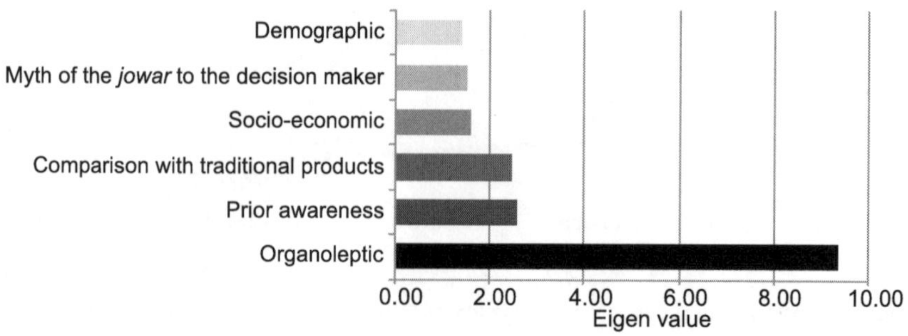

Fig. 6.21: Factors for consideration in *jowar suji*/*rawa*/semolina

Clusters of the respondents

Clusters of the consumers based on response on purchase decision are presented in Table 6.16.

Table 6.16: Clusters of the consumers based on response on purchase decision

Cluster	*Purchase*	*Not Purchase*	*Total Respondents*	*% Purchase*	*Cluster Output*
1	250	28	278	90	Average
2	206	3	209	98	Good
3	11	41	52	21	Bad
4	6	18	24	25	Bad
5	167	10	177	90	Good

The *jowar suji/rawa*/semolina clusters are significantly different in terms of price level, place of purchase, buying preferences (branded or unbranded), per cent expenditure on food, monthly household income, occupation of the female members of the family, and number of children (Figs 6.22–6.24). Details of the consumer profile from both good and average cluster are given in the Annexure.

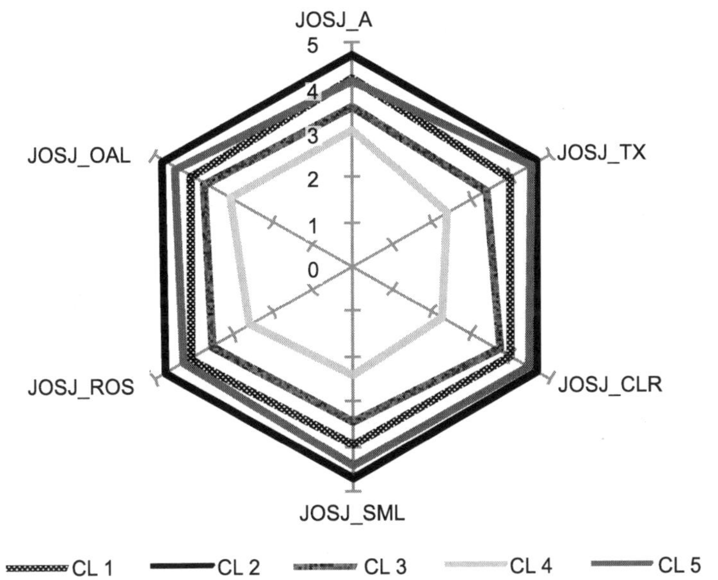

(Score: 5 = Liked very much, 1 = Not liked at all)

Fig. 6.22: Score on the attributes of *jowar suji/rawa*/semolina (raw), by cluster

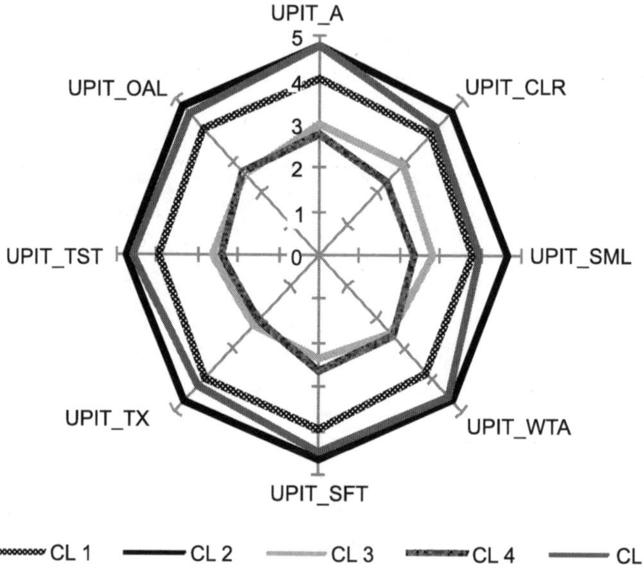

(Score: 5 = Liked very much, 1 = Not liked at all)

Fig. 6.23: Score on the attributes of *jowar suji/rawa*/semolina attributes (*uppit*), by cluster

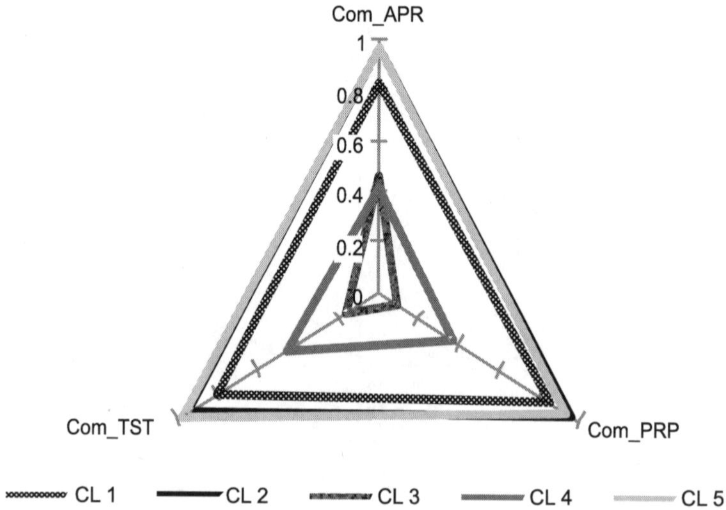

(Score: 0 = Traditional product is better, 1 = Jowar is better or both are same)

Fig. 6.24: Score on the comparison of attributes of *jowar suji/rawa*/semolina with traditional *suji/rawa*/semolina, by cluster

Attributes influencing the purchase of jowar suji/rawa/ semolina (uppit)

Ease and similarity in preparation of *jowar rawa* when compared with wheat *rawa* is a very important factor/attribute. The respondents seek no variation in the preparation process that they follow for making *rawa*. Taste of *jowar rawa* in comparison with the wheat *rawa* is the second important attribute influencing the purchase decision. As shown in Table 6.17, these two attributes could increase the purchase of *jowar rawa/suji* by 3.77 and 2.67 times, respectively, when there is an increase of 1 unit in liking scale or if the respondents find the *jowar rawa* to be similar as the regular *rawa* used by them. Taste of the final product (*uppit*) is another important attribute that could increase the purchasing chance by 1.99 times followed by the *uppit* appearance increasing the purchase by 1.84 times.

Table 6.17: Odds Ratio Estimate to predict the attributes influencing the purchase of *jowar suji/rawa*/semolina

Factor	*MLE Estimate*	*Odds Ratio Estimate*
Uppit appearance	0.6117	1.844
Uppit smell	0.4439	1.559
Uppit taste	0.6917	1.997
Preparation in comparison with wheat *rawa*	1.3283	3.775
Taste in comparison with wheat *rawa*	0.9826	2.671

Note: The analysis of Maximum Likelihood Estimates (MLE) was used to obtain parameter estimates. Significance of parameter estimates was based on the Wald Chi-square value at $P < 0.05$.

Prediction of correct purchase decision

To draw an inference, 631 or 85 per cent of the total respondents who are willing to purchase and 61, or 8 per cent of the consumer respondents who are not interested in purchasing the *jowar rawa* are observed and predicted as correct classification group. On the other hand, 39, or 5 per cent of the total consumer respondents who are willing to purchase and 9 or 1 per cent of the consumer respondents who are not interested in purchasing the *jowar rawa* are observed and predicted as incorrect classification group. Thus, the purchasing decision of the *jowar rawa* could be predicted with 94 per cent of accuracy and 6 per cent of inaccuracy (Table 6.18).

Table 6.18: Correct classification of actual purchase decision and prediction on correctness of purchase for *jowar suji/rawa/* semolina

	No. of Respondents			*% of Respondents*		
	Not Purchase	*Purchase*	*Total*	*Not Purchase*	*Purchase*	*Total*
Not Purchase	61	39	100	8%	5%	14%
Purchase	9	631	640	1%	85%	86%
Total	70	670	740	9%	91%	100%
Correct Prediction			692	Correct Prediction		94%
Incorrect Prediction			48	Incorrect Prediction		6%

Purchase prediction accuracy based on individual attributes (hit ratio)

Taste of the final product of *rawa* (*uppit*) can be used as a single predictor, and prediction of the purchase decision of *jowar rawa* can be done with accuracy of 73.7 per cent. Smell and appearance of the final product (*uppit*) are also important attributes affecting the purchase decision of the consumer that can be predicted with 72.5 per cent and 70.7 per cent accuracy, respectively. Similarity in the preparation process and taste to that of traditional *rawa* can also be used to predict the purchase decision with 59.9 per cent and 57.5 per cent accuracy, respectively (Fig. 6.25).

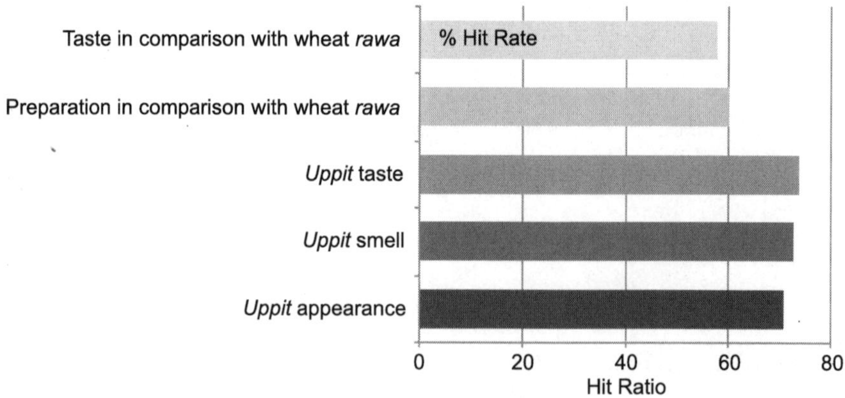

Fig. 6.25: Purchase prediction accuracy based on individual attributes for *jowar rawa* (*uppit*)

Similarities and dissimilarities in the response of the respondents on important attributes of jowar suji/rawa/semolina

The comparison here is done between the two cities of Hyderabad and Pune as Delhi respondents did not prepare *uppit* from the *rawa*. ANOVA analysis could not be done due to the absence of pair-forming variables.

Hyderabad and Pune have significant difference in the responses of appearance, smell, and taste of the final *rawa* product (*uppit*). The smell of the *uppit* made from *jowar* is not liked by the respondents of Pune (Tables 6.19 and 6.20).

Table 6.19: Response on the influential attributes of *jowar suji*/*rawa*/semolina, by city

	Hyderabad	*Pune*	*t-Value*	*Ps > (t)*
Sample Size	434	305		
Mean UPIT_A	4.4309	4.1508	4.76	< .0001
Mean UPIT_SML	4.0624	3.8492	3.03	.0025
Mean UPIT_TST	4.3341	4.0197	5.15	< .0001

Table 6.20: Response on the influential attributes of *jowar*, by city vis-à-vis wheat *rawa*

	Hyderabad		*Pune*		*Chi-square Test*	*Probability*
	0	*1*	*0*	*1*		
Taste in comparison with wheat *rawa*	60	375	69	236	9.71	0.0018
Preparation in comparison with wheat *rawa*	49	386	76	229	23.8	< .0001

Sira/Sweet/*Halwa*

The value for money and food habits constituted the last factor having a value of 1.12. The reliability of the data and consistency in the responses received for Factor 1 pertain to 0.940, which is much higher than the measuring point of 0.600, also higher as compared to Factors 2 and 3, which thus can be explained as a more reliable response. The per cent variance explained by these six factors is 57.5 per cent (Table 6.21; Fig. 6.26).

Table 6.21: Important factors/dimensions of the response of consumers on *suji*/*rawa*/ semolina

Dimensions	*Cronbach's Alpha*	*Eigen Value*	*Variance Explained (%)*
Organoleptic	0.940	8.13	25
Prior awareness	0.917	4.91	15
Comparison with traditional products	0.641	1.68	5
Socio-economic	N.A.	1.64	5
Demography	N.A.	1.38	4.0
Value for money and food habits	N.A.	1.12	3.5

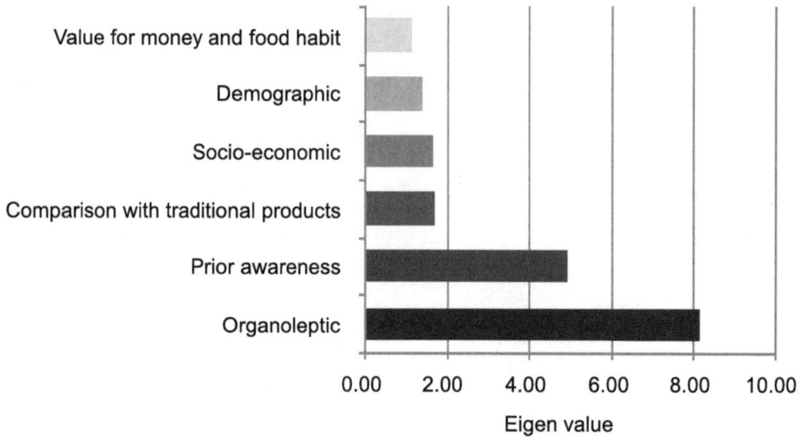

Fig. 6.26: Factors for consideration in *jowar rawa* (*sira*)

Clusters of the respondents

The *jowar suji*/*rawa*/semolina clusters for *sira* are widely different from the *uppit* clusters. The clusters have more similarities among the respondents. Place of purchase, buying preferences (branded or unbranded), quantity of *rawa* consumed in a month, and number of children in a family or household are the major parameters that determine significant differences between good cluster and average cluster (Table 6.22; Figs 6.27–6.29). Details of the consumer profile from both good and average clusters are given in the Annexure.

Table 6.22: Clusters of the consumers based on the response on purchase decision

Cluster	*Purchase*	*Not Purchase*	*Total Respondents*	*% Purchase*	*Cluster Output*
1	1	27	28	3.57	Bad
2	81	56	138	59	Bad
3	159	35	194	81	Average
4	101	13	114	89	Good

Attributes influencing the purchase of jowar suji/rawa/semolina (sira)

The analysis of maximum likelihood estimates (MLE) was used to obtain parameter estimates. Significance of parameter estimates was based on the Wald Chi-square value at $P < 0.05$.

Attributes of *jowar rawa* in comparison with wheat *rawa* in terms of appearance, preparation process, and taste are the major factors influencing

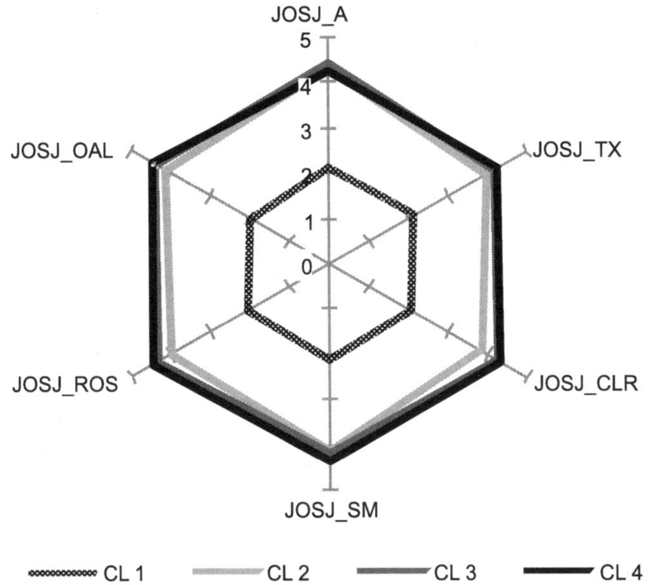

(Score: 5 = Liked very much, 1 = Not liked at all)

Fig. 6.27: Score on *jowar rawa* attributes, by cluster

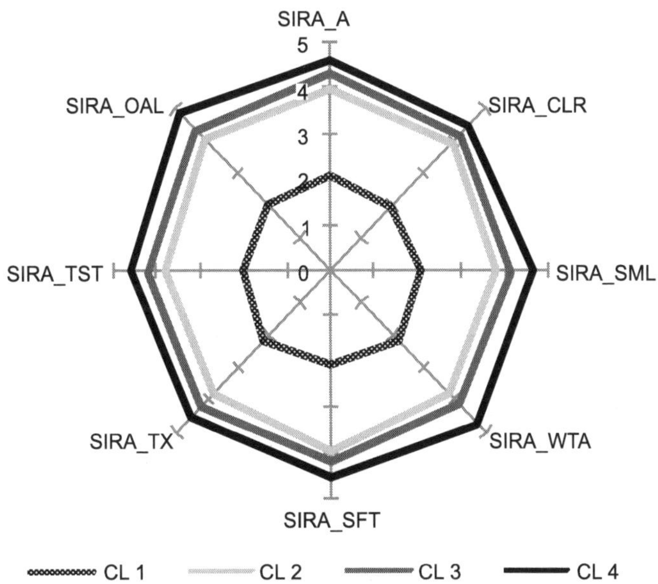

(Score: 5 = Liked very much, 1 = Not liked at all)

Fig. 6.28: Score on *jowar rawa* attributes, by cluster

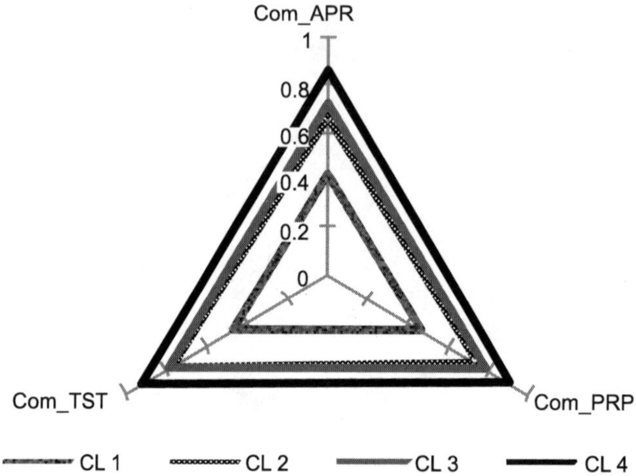

(Score: 0 = Traditional product is better, 1 = Jowar is better or both are same)

Fig. 6.29: Preference of *jowar rawa* attributes in comparison with traditional, by cluster

the purchasing decision of the consumers. As shown in Table 6.23, these three attributes could increase the purchasing chance of *jowar rawa/suji* by 4.21, 3.19, and 4.57 times, respectively, when there is an increase in 1 unit in the liking scale or if these are similar to the attributes of wheat *rawa*. Smell of the final product (*sira*) is another important attribute that could increase the purchasing chance by 1.77 times. Colour of *jowar rawa* is another important attribute influencing the first sight purchasing decision followed by the texture of the *rawa*. Both attributes could increase the purchasing chances by 2.24 and 1.47 times, respectively.

Table 6.23: Odds Ratio Estimate to predict attributes influencing the purchase of *jowar suji/rawa*/semolina

Factor	*MLE Estimate*	*Odds Ratio Estimate*
JOSJ_TX	0.3834	1.467
JOSJ_CLR	0.8098	2.247
SIRA_SML	0.5719	1.772
Com_APR	1.4385	4.214
Com_PRP	1.1588	3.186
Com_TST	1.5201	4.573

Prediction of correct purchase decision

Thus, 316, or 67 per cent of the total consumer respondents who are willing to purchase and 71 or 15 per cent of the consumer respondents who are not interested in purchasing the *jowar rawa* are observed and predicted as correct classification group. On the other hand, 58 or 12 per cent of the total consumer respondents who are willing to purchase and 29 or 6 per cent of the consumer respondents who are not interested in purchasing the *jowar rawa* are observed and predicted as incorrect classification group. In conclusion, the purchasing decision of *jowar rawa* could be predicted with 82 per cent accuracy and 18 per cent of inaccuracy (Table 6.24).

Table 6.24: Correct classification of actual purchase decision and prediction on correctness of purchase for *jowar suji/rawa*/semolina (*sira*)

	No. of Respondents			*% of Respondents*		
	Not Purchase	*Purchase*	*Total*	*Not Purchase*	*Purchase*	*Total*
Not Purchase	71	58	129	15%	12%	27%
Purchase	29	316	345	6%	67%	73%
Total	100	374	474	21%	79%	100%
Correct Prediction			387	Correct Prediction		82%
Incorrect Prediction			87	Incorrect Prediction		18%

Purchase prediction accuracy based on individual attributes (hit ratio)

Purchase prediction accuracy based on individual attributes for *jowar rawa* (*sira*) are given in Fig. 6.30.

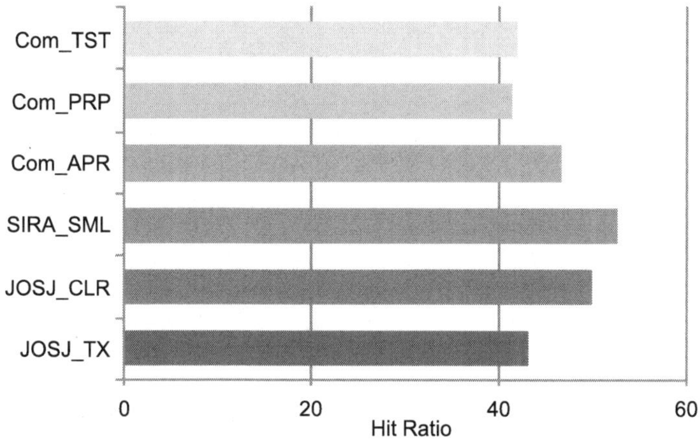

Fig. 6.30: Purchase prediction accuracy based on individual attributes for *jowar rawa* (*sira*)

Similarities and dissimilarities in the response of the respondents on important attributes of jowar suji/rawa/semolina (sira)

Delhi respondents moderately liked the important attributes of *jowar rawa* while preparing *sira* (*halwa*) from it, and liking of the respondents from Pune was average (Table 6.25).

Table 6.25: Response on influential attributes of *jowar suji/rawa/*semolina (*sira*), by city

	Delhi	*Pune*	*t-Value*	*Pr > (t)*
Sample	373	100	NA	NA
JOSJ_CLR	4.0188	3.96	0.65	0.514
Texture	4.0241	3.93	1.13	0.2587
SIRA_SML	4.131	3.75	3.99	< .0001

Delhi and Pune respondents are very close in terms of their perception on *jowar rawa*'s colour and texture, whereas on the smell of *sira* or the final product, response in both cities is significantly different. Similarly, in comparison with traditional *rawa*, response of both the cities on *jowar rawa* is not significantly different from each other (Table 6.26).

Table 6.26: Response on influential attributes of *jowar rawa* (*sira*), by city

	Delhi		*Pune*		*Chi-square Test*	*Probabilty*
	0	*1*	*0*	*1*		
Appearance in comparison with wheat *rawa*	109	265	20	80	3.33	0.068
Preparation in comparison with wheat *rawa*	77	297	28	72	2.51	0.1129
Taste in comparison with wheat *rawa*	76	298	20	80	0.005	0.943

(Score: 0 = Traditional rawa is better, 1 = Jowar is better or both are same)

Jowar Flakes

The reliability of the data and consistency in the responses received for Factor 1 pertain to 0.952, which is much higher than the measuring point of 0.600. Cronbach alpha for Factors 2, 3, and 4 is 0.914, 0.732, and 0.572 respectively. The per cent variance explained by these six factors is 61.5 per cent (Table 6.27; Fig. 6.31).

Table 6.27: Important factors/dimensions of the consumer's response on *jowar* flakes

Dimensions	*Cronbach's Alpha*	*Eigen Value*	*Variance Explained (%)*
Organoleptic	0.952	8.44	28
Prior awareness	0.914	4.20	14
Myths associated with *jowar*	0.732	1.89	6
Comparison with traditional products	0.572	1.53	5
Socio-economic and value for money	N.A.	1.40	4.5
Demography	N.A.	1.27	4.0

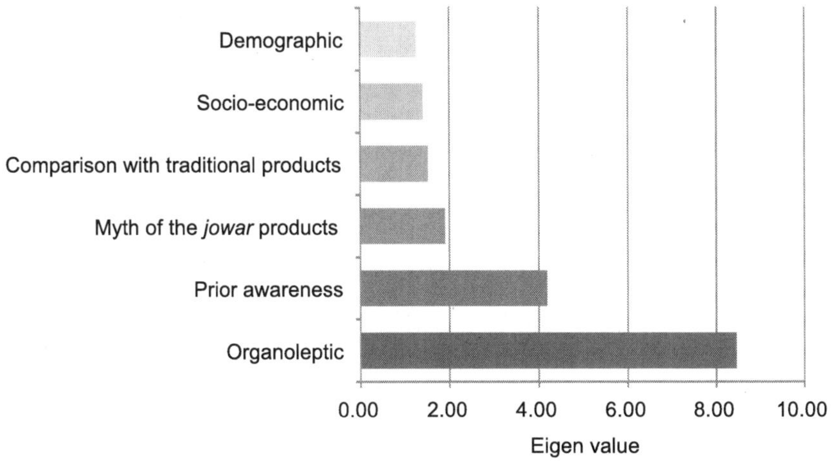

Fig. 6.31: Factors for consideration in *jowar* flakes

Clusters of the respondents

Socio-economic profile of the respondents grouped into good and average clusters is presented in Table 6.28 and Figs 6.32–6.34. Most of the parameters, except occupation of the male member of the family and family size, show significant difference between good and average clusters. Details of the consumer profile from both good and average clusters are given in the Annexure.

Table 6.28: Clusters of the consumers based on the response on purchase decision

Cluster	*Purchase*	*Not Purchase*	*Total Respondents*	*% Purchase*	*Cluster Output*
1	105	4	109	96	Good
2	521	63	584	89	Good
3	173	57	230	75	Average
4	0	34	34	0	Bad
5	146	112	258	56	Bad

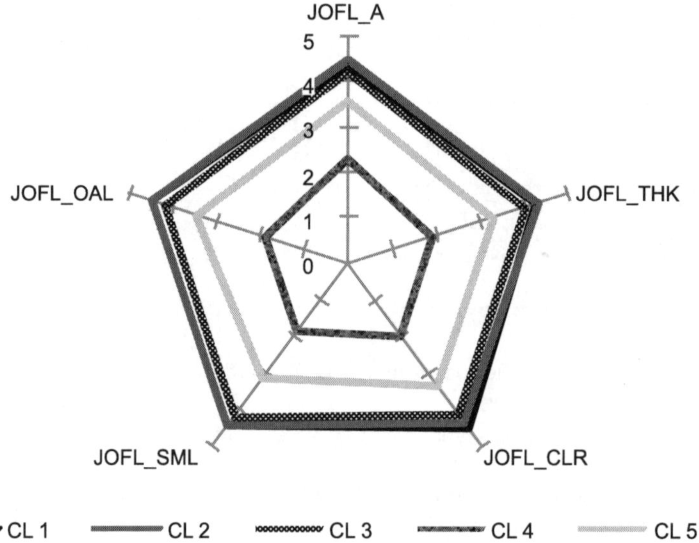

(Score: 5 = Liked very much, 1 = Not liked at all)

Fig. 6.32: Score on the attributes of raw *jowar* flakes, by cluster

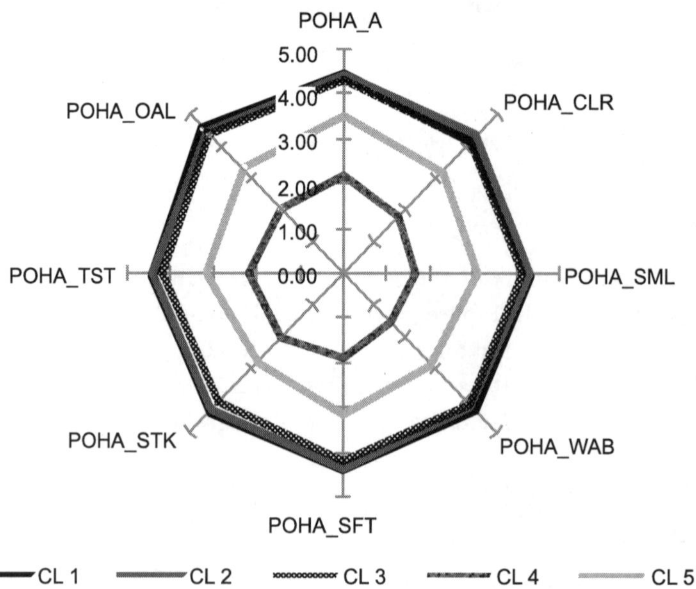

(Score: 5 = Liked very much, 1 = Not liked at all)

Fig. 6.33: Score on the attributes of cooked *jowar* flakes (*poha*), by cluster

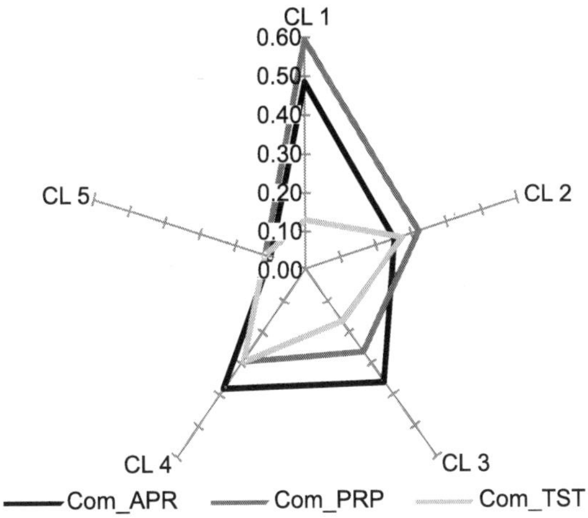

(Score: 0 = Traditional product is better, 1 = Jowar is better or both are same)

Fig. 6.34: Score on the comparison of the attributes of *jowar* flakes with traditional flakes, by cluster

Attributes influencing the purchase of jowar flakes

Perception about the *jowar* product as an energy source could drive the chances of purchase by 2.74 times. In other words, the product can be marketed as a health product alone, like *Kellogg's* breakfast cereals. Similarity in the preparation process of *jowar* flakes to that of rice flakes can alone increase the chance of the purchase of the former by 2.20 times with the increase in 1 unit of the liking scale. In the similar way, appearance, water absorption, softness, and overall experience of cooked flakes (*poha*) could increase the purchasing chance by 1.75, 1.79, 1.58, and 1.63 times, respectively, if there is 1 unit increase in liking scale (Table 6.29).

Table 6.29: Odds Ratio Estimate to predict the attributes influencing the purchase of *jowar* flakes

Factor	*MLE Estimate*	*Odds Ratio Estimate*
Stat_7	1.0094	2.744
POHA_A	0.5612	1.753
POHA_WAB	0.5829	1.791
POHA_SFT	0.458	1.581
POHA_OAL	0.49	1.632
Com_PRP	0.7902	2.204

Note: The analysis of maximum likelihood estimates (MLE) was used to obtain parameter estimates. Significance of parameter estimates was based on the Wald Chi-square value at $P < 0.05$.

Prediction of correct purchase decision

As shown in Table 6.30, 885, or 73 per cent of the total respondents, who are willing to purchase, and 140, or 12 per cent of the respondents, who are not interested in purchasing *jowar* flakes, are observed and predicted as the correct classification group. On the other hand, 130, or 11 per cent of the total respondents who are willing to purchase and 60, or 5 per cent of the respondents, who are not interested in purchasing the *jowar* flakes are observed and predicted as the incorrect classification group. In conclusion, the purchasing decision of *jowar* flakes could be predicted with 84 per cent accuracy and 16 per cent inaccuracy.

Table 6.30: Correct classification of actual purchase decision and prediction on the correctness of purchase for *jowar* flakes

	No. of Respondents			*% of Respondents*		
	Not Purchase	*Purchase*	*Total*	*Not Purchase*	*Purchase*	*Total*
Not Purchase	140	130	270	12%	11%	22%
Purchase	60	885	945	5%	73%	78%
Total	200	1015	1215	16%	84%	100%
Correct Prediction			1025	Correct Prediction		84%
Incorrect Prediction			190	Incorrect Prediction		16%

Purchase prediction accuracy based on individual attributes (hit ratio)

Fig. 6.35 gives the purchase prediction accuracy based on individual attributes for *jowar* flakes

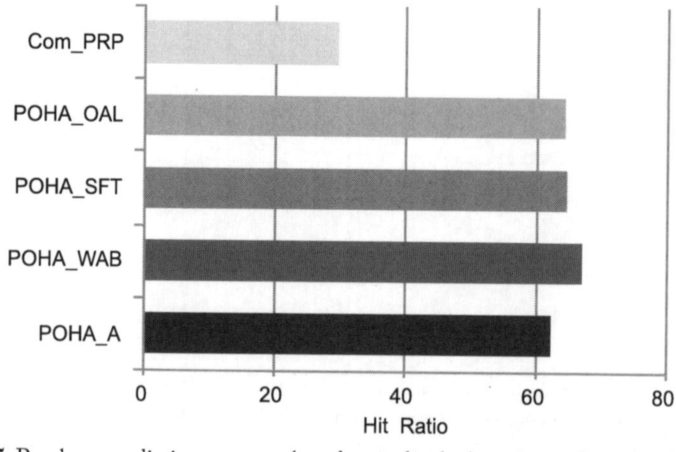

Fig. 6.35: Purchase prediction accuracy based on individual attributes for *jowar* flakes

Similarities and dissimilarities in the response of the respondents on important attributes of jowar flakes

Hyderabad is symbolised as City 1, Delhi as City 2, and Pune as City 3. Delhi and Hyderabad have similar response in terms of the appearance of the cooked *jowar* flakes (*poha*), overall experience, water absorption of *poha*, and its softness attributes. Apart from these, the response of the rest two pairs of the cities is significantly different in terms of mentioned attributes (Fig. 6.36).

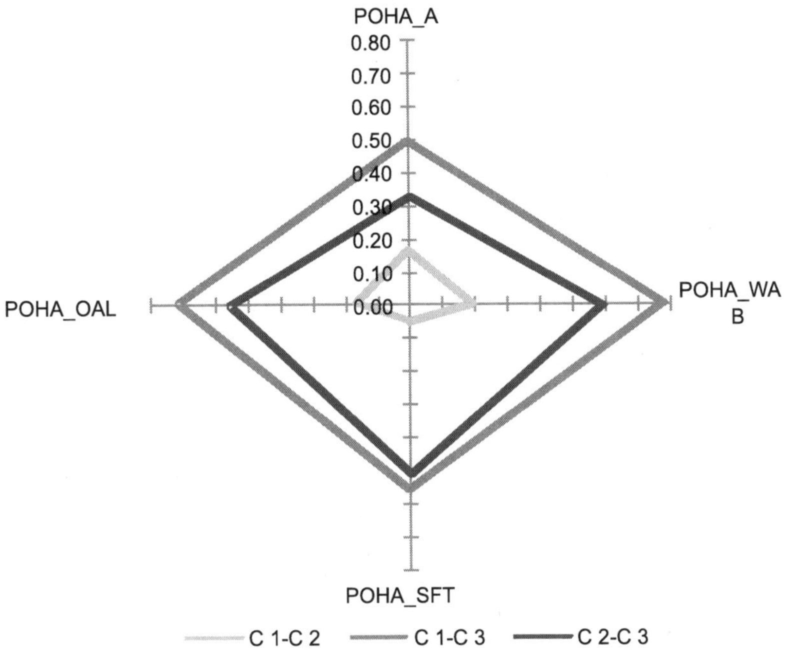

(Scale: 1 = Jowar product is better or both are same, 0 = Traditional product is better)

Fig. 6.36: Pair wise response of cities, on influential attributes of *jowar* flakes

In Delhi and Hyderabad, respondents liked to a larger extent the important attributes of *jowar* flakes influencing purchase decision. The average score for these attributes is above 4.15, which is near to the extreme liking of the product. In Pune, respondents are still hesitant in replacing the rice flakes with *jowar* flakes considering the attributes mentioned above. The average score assigned by the respondents is near to moderately liking the product (Fig. 6.37).

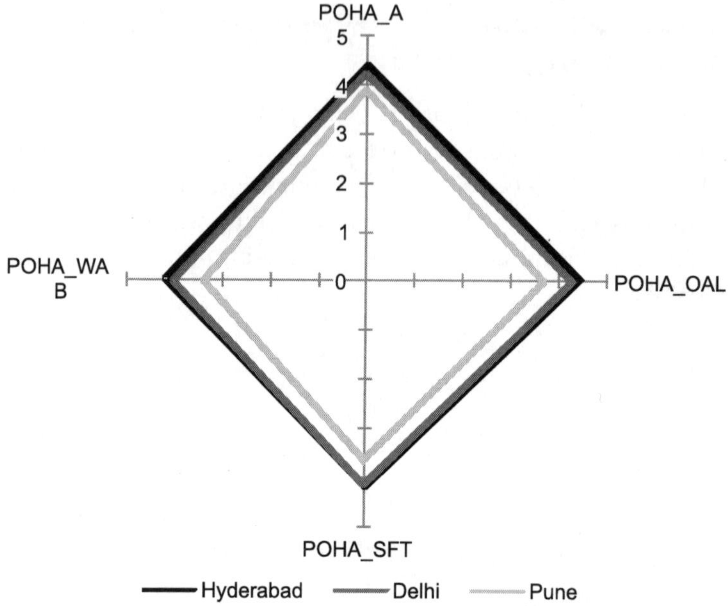

(*Scale*: *5 = Liked very much, 1 = Not liked at all*)

Fig. 6.37: Analysis of the response of individual city on important attributes

CHAPTER 7

Public-Private Partnership Opportunities

Public-Private Partnership Opportunities in Developing and Marketing Sorghum-based RTC/RTE Products

First initiative on the public-private partnership (PPP) in a big way happened when Directorate of Sorghum Research (DSR) linked MoU with ITC (ABD) in 2008 under the National Agricultural Innovation Project (NAIP). Many other linkages across the sorghum-growing belts are occurring after the government launched the programme Initiative for Nutritional Security through Intensive Millets Promotion (INSIMP) in 2011/12. Backward integration of sorghum value chain was pilot tested by ITC (ABD) Ltd under the DSR-led consortium on Millets Value Chain, funded by NAIP/Indian Council of Agricultural Research (ICAR) (World Bank aided). This was enabled with the support of *e-choupal* platform where beneficiary farmers are market assured on the functionality, feasibility, and sustainability of intensive sorghum cultivation. It is a relatively new phenomenon in Indian millet scenario, which has paved the way for supply chain management and has further strengthened the development and commercialisation of sorghum products. Further, the product with specific farm production led to the procurement, aggregation, and linking with other stakeholders in the value chain, thus ultimately connecting the producers

with consumers. Though this sorghum value chain existed earlier also in a disintegrated manner. Since the fair average quality (FAQ) grains and their quantity were predetermined under the PPP, both the parties could rely on each other; this provided a better price to the farmers and sufficient quantity and desired quality of sorghum to the processors for smooth functioning of their business.

End product specific cultivars developed/identified by DSR are technologically back stepped to all the beneficiary farmers in Maharashtra and Andhra Pradesh. In this respect, the ITC (ABD) Ltd for the first time took up farm extension services in sorghum on pilot scale along with a buyback arrangement. The identity preserved grain was procured and bulked to overcome supply chain issues under low marketed surplus regime.

Developing sorghum products *vis-à-vis* determining their nutritional values, and simultaneously promoting and marketing them have enhanced its awareness preference in recent years. Rural *chaupal haats* were arranged to create awareness on this new initiative. The value chain approach thus provided end-to-end solutions paving the way for forward integration or processing and commercialisation by the small and medium scale entrepreneurs. The impact of this successful PPP collaboration has enhanced the productivity and resulted in higher incomes for the stakeholders. There was an increased allocation of better lands for sorghum cultivation, which can be attributed to the impact of the PPP initiative.

Among other initiatives worth mentioning is the MoU entered between DSR and M/s Britannia Industries Ltd, Bengaluru, a multinational company in food sector, to undertake Joint R&D on sorghum biscuits and bakery products for three years, wherein the DSR will be identifying cultivars suitable for sorghum-based products and pilot test backward integration while Britannia will come up with some sorghum-based products with their functional portfolio and rheological properties studied in depth by both the partners. This PPP is expected to lead to create demand for sorghum as the uptake is going to be in large volumes and the primary stakeholder may be benefitted in the long run. Britannia Industries is expected to categorise the sorghum-based products under their Nutri-choice banner.

Status of sorghum processing and commercialisation

Primary processing of sorghum is necessary to counter the inconveniences in sorghum preparation and consumption. Secondary processing has begun with the increase in the awareness on sorghum's nutritional values by the consumers. With the implementation of INSIMP project initiated by Department of Agriculture and Cooperation(DAC), Government of India, a mass pilot-scale sorghum processing is taking place across the sorghum-growing states.

Sorghum processing technologies have been developed in the country with DSR playing the key role in developing them while promoting and commercialising them as health food meant for all age groups and niche segments such as diabetic patients, elderly, and school-going children. The knowledge has also been disseminated to the farmers through entrepreneurship development programmes so that they are either aware of the various developments that have been taking place or they learn to start their own processing at the farm level.

Since there is no machinery in the market exclusively meant for sorghum processing, the available machineries meant for processing other fine cereals such as wheat and rice are procured and retrofitted for sorghum processing.

Marketing opportunities and its sustainability

Based on the outcome of the consumer study, market opportunities for sorghum products and core strategies to be added in marketing of these products are explained in onward sections.

Branding, targeting, and positioning

Sorghum products are complete family package, promising wholesome nutrition for all members of the family, perfectly substitutable with conventional grains, as there is no difference in terms of preparation, appearance, and taste. These products are also a convenient food as no extra effort or time is required for the preparation. All products are positioned as 'healthy nutri-cereal' products.

Product branding strategy

Consumers have different perception for the various types of products they consume. In case of sorghum products the study has recommended different target strategies to be adopted for the marketing of these products.

Jowar-rich multigrain atta

- Healthy not at the cost of taste: '*Sehat Banao, Swad bhi Pao*'
- Trustable and of good quality: Certified by National Institute of Nutrition (NIN)
- Diet management: Very filling in small quantities
- Large scope with respect to the variety of products
- Tag line: 'Homemade or food from ancestor's bag'

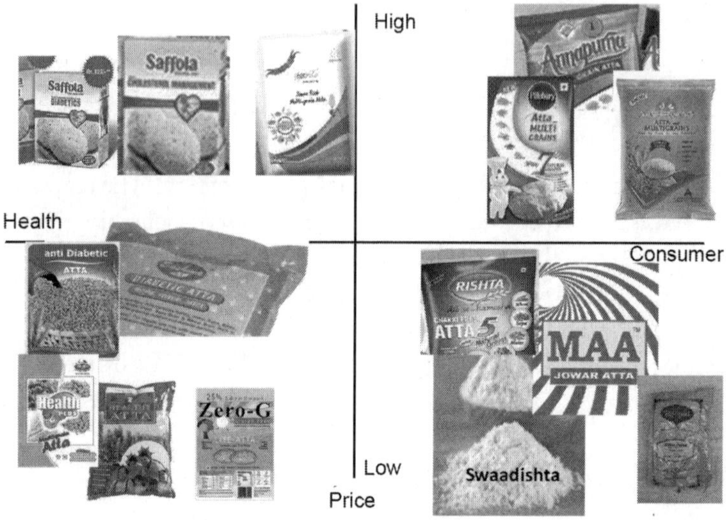

Fig. 7.1: Placement/positioning of *jowar*-rich multigrain *atta*

Note: The pure sorghum flour is positioned in consumer segment and lower price level while the multigrain *atta* also fall in consumer segment but it is high priced. On the other hand, when we see the products which are positioned as healthy, national brand like 'Saffola' is high priced as compared to the regional brands or label within the similar segment. After the consumer perception analysis, the study proposes to position the '*Jowar*-rich Multigrain *Atta*' in the healthy segment with price equivalent of the national branded products.

Biscuit branding strategy

- Healthy not at the cost of taste: '*Sehat Banao, Swad bhi Pao*'
- Trustable and of good quality: Certified by National Institute of Nutrition (NIN)
- Diet management: Sufficient in small quantities

Segmentation

- Household of family comprising four to five adult members.
- The household may have no child or a single child of less than 12 years of age.
- Married women (mostly housewives).
- B2-A2 class of income group (monthly household income of Rs. 11,000-25,000).
- Important pockets in cities like Hyderabad and Pune that consume traditional food.

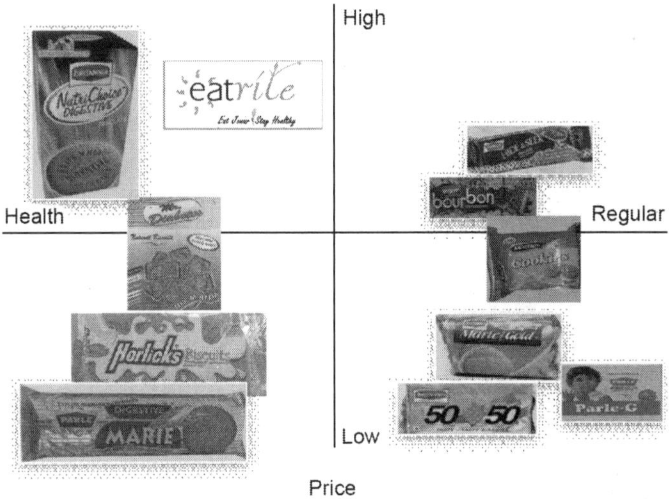

Fig. 7.2: Placement/positioning of *jowar* biscuits

Note: Biscuits currently available in the market are either positioned as regular consuming or as healthy consuming. The products shown in Figure 7.2 explain the positioning of the various biscuits and their pricing. The study recommends positioning of *jowar* biscuits in the healthy segment and its pricing should be at par with other biscuits in higher price position.

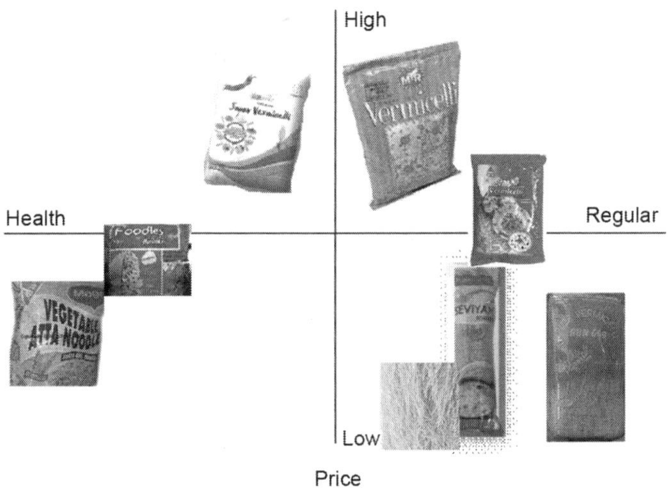

Fig. 7.3: Placement/positioning of *jowar* vermicelli

Note: In the vermicelli market regional brands dominate and the pricing is again dependent on the goodwill of the brand (Figure 7.3). Currently in the healthy segment, instant needles like products are available and in terms of pricing they are priced moderately. *Jowar* vermicelli is advised to be positioned at the healthy product segment and in high price position.

- **Hyderabad**: Kukatpalli, Begum Bazaar, Red Hills, Gaddi Annaram, Old Malakpet, Musheerabad, Sanathnagar (total population about 3 lakh; 1.74 kg/year per capita consumption = 529 tonnes)
- **Pune:** All *peth* areas, Shivajinagar, Bibewadi, Shankar Seth road, and Bund Graden (total population about 3.5 lakh; 10.17 kg/year per capita consumption = 3560 tonnes)

Targeting and positioning

- Complete family product: Suitable for individuals, family, young ones, elderly, kids, health conscious members, and traditional members.
- Healthy product: Provides wholesome nutrition for all members of the family; it is a myth that it is only for diabetic/elderly/sick members; good source of energy powerhouse; useful for diet management.
- Perfectly substitutable: Not very dissimilar to regular *atta* in terms of preparation, appearance, and taste.
- Convenience food: No extra effort or time required for preparation
- Addresses both health and taste issues.
  - Available products are either positioned as 'health' (cholesterol management and diabetic-friendly [Saffola], gluten-free) or 'taste' (Ashirvaad, Pillsbury, Annapurna, etc.) food.
  - Available biscuits are either positioned as 'health' (digestive, light, nutritious, diabetic-friendly) or 'taste' (Good Day, Hide and Seek, Butter Bite or bakery (nan-khatai, fresh baked) food

Marketing, advertisement, and communication

There is a need to create awareness and launch media campaign to popularise the products so as to make the sorghum-based products available to the masses. Road shows; establishing display booths at malls/super markets, residential colonies, gym, and health centres; using visual media; enlisting the services of brand ambassadors and opinion leaders; organising events like cooking competitions and food festivals are effective tools of marketing.

Suggestions on the marketing and advertisement

- Brand promotion: Advertising through celebrity, preferably sports person (using the concept of the importance of the product both in the field (source of energy) and at home (delicacy).

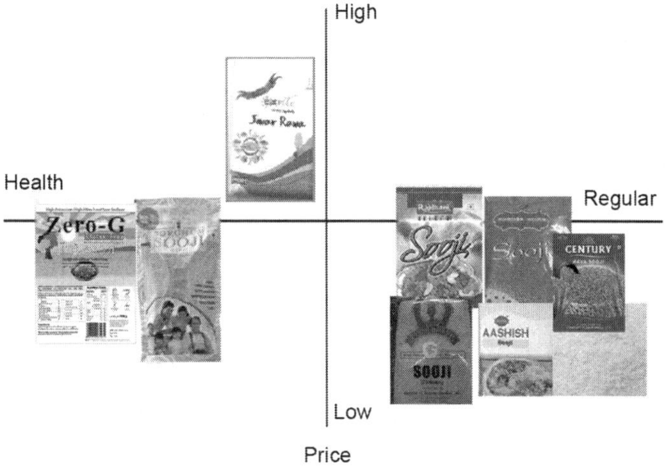

Fig. 7.4: Placement/positioning of *jowar suji/rawa*/semolina

Note: Currently very few manufacturers are supplying *jowar* made *suji/rawa*/semolina in the market. Many of these products are sold online rather than being placed in the retail stores. The jowar raws manufactured by DSR is recommended to be placed in the healthy segment with a higher price as compared to the regular *suji/rawa*/semolina available in the market.

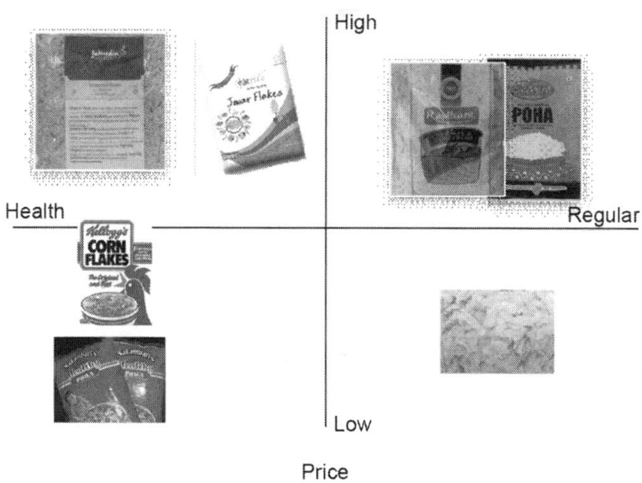

Fig. 7.5: Placement/positioning of *jowar* flakes

Note: As shown in Figure 7.5, within the healthy flake segment, currently corn flakes are the only products available in the market. The private label corn flakes are sold at lower price as compared to the national level brands. Similarly some organic jowar flake is also available in the market but currently sold to the niche segment. The studies recommend that jowar flakes should be placed in the healthy segment and its price should be higher as compared to the corn flakes.

- Awareness campaign: Road shows and establishing display booths at mall/super markets, residential colonies, gymnasium, and health centres.
- Media campaign:
  - Visual Communication; enlisting the services of brand ambassador and opinion leaders; organising cooking competition and food festivals, etc.
  - Advertisement in newspapers (regional and national), health magazines, TV, and FM radio.

Production and sales

Decentralisation of production is necessary ascatering to the markets like Delhi with the centralised operation from Hyderabad would not be economically feasible. Even in the initial stages, it would hurt the profitability to a large extent, as market entry costs are huge.

Target market selection

Market scenario analysis is one of the important tools to formulate the marketing strategy before launching any new products in the market. In case of *jowar* based products, the surveyed cities have been described as per important market dynamics. These would be useful to determine the appropriate marketing strategies.

Table 7.1: Target market selection for *jowar*-rich multigrain *atta*

	Hyderabad	*Pune*	*Delhi*
Classification of the city	A 1	A	A 1
Respondent's classification	Early majority	Late majority	Innovators
Awareness on the benefits of *jowar*	High	High	Low
Per capita expenditure on food	Medium	Low	High
Branded *atta* buyer	Branded	Unbranded	Branded
Overall liking of the product	Very good	Good	Very good
Annual consumption demand of wheat *atta* (000' tonne)*	78.6	152	1402
Size of multigrain *atta* (tonne)	7.86	15.2	28.04
Opportunity space (1%) (tonne)	160	150	280
Expected price (Rs/kg)	*16–25*	*15–20*	*18–21*

* Estimate: NSSO data for per capita consumption of selected food item

Delhi being a non-traditional consumer of *jowar* appears as one of the largest market (in terms of volume) for *jowar* -rich multigrain *atta*. On the other hand, Hyderabad a traditional consumer of *jowar*, shows substantial opportunity space for *jowar* -rich multigrain *atta*. Also the expected price range to be paid by the consumer is higher.

Table 7.2: Target market selection for *jowar* vermicelli

	Hyderabad	*Pune*	*Delhi*
Annual demand of vermicelli (000' tonne)*	2.8	0.5	3.6
Opportunity space (1%) tonne	28	5	36
Expected price (Rs/200 g)	11-14	10-12	15-18

*Estimate: NSSO data for per capita consumption of selected food item

We have assumed that out of the market size of vermicelli in the surveyed region, 1 per cent of the total demand can be replaced with *jowar* vermicelli. The expected price of *jowar* vermicelli which consumers are willing to pay is higher in Delhi market. In terms of the market size, Hyderabad has a huge opportunity space being a traditional consumer of *jowar*.

Table 7.3: Target market selection for *jowar* biscuits

	Hyderabad	*Pune*	*Delhi*
Annual demand of biscuits (000' tonne)*	8.3	11.8	37.3
Opportunity space (tonne) (0.5%)	40	59	186
Expected Price (Rs/100 g)	6-9	5-6	8-10

* Estimate: NSSO data for per capita consumption of selected food item

Consumers of Hyderabad and Pune cities have said that their expected buying price of *jowar* biscuits would be ranging from Rs. 5-9 for the pack of 100 grams. This price range is much lower than the expected purchase price of Delhi consumers. In terms of market volume Delhi is showing larger opportunity space as compared to Hyderabad and Pune market.

Table 7.4: Target market selection for *jowar suji*/*rawa*/semolina

	Hyderabad	*Pune*	*Delhi*
Annual demand of wheat *rawa* (000' tonne)*	30	4.5	7.6
Opportunity space (tonne) (1 %)	300	45	76
Expected price (Rs/kg)	20-25	20-25	20-25

* Estimate: NSSO data for per capita consumption of selected food item

Hyderabad has largest opportunity space (market size in terms of volume) for *jowar suji*/*rawa*/semolina. The consumer's expected purchase price of *jowar rawa* is similar for all the cities.

Table 7.5: Target market selection for *jowar* flakes

	Hyderabad	*Pune*	*Delhi*
Annual average demand of rice flake (000' tonne)*	1.9	5.0	11.8
Opportunity space (tonne) (1%)	19	50	118
Expected price (Rs/kg)	15–20	18–20	18–22

* Estimate: NSSO data for per capita consumption of selected food item

A rice flake is one of the most popular snack foods in the northern and western India. Consumer's expected purchase price for *jowar* flakes in Delhi and Pune cities is almost similar.

Marketing mix: Four P's of marketing

Product

Jowar products are meant for all. It is not only healthy but also as tasty as regular products like *atta*, vermicelli, *rawa*, biscuits, and flakes. The attributes of *jowar* products are not different from those of the regular/traditional products in terms of appearance in raw form, preparation process/methods, and taste in the final form. *Jowar* products should not be placed as high 'premium products' but should carry health tags.

Convenience could be key feature of the *jowar* products in traditional consuming areas and urban centres like Hyderabad and Pune. However, factors like preparation difficulties and limited product options of *jowar* have brought down the demand of *jowar*. Versatile product range of *jowar* would reinforce the consumption in the urban areas.

Price

Price of the product should be at par with the other similar health products available in the market. 'Skim the cream' could be the suggested pricing strategy for the healthy products, but in the place like Pune, discounted pricing or introductory price is suggested. Cost-based pricing strategy could also be adopted, as the consumer's expectations with respect to prices are conservative.

Promotion

A high level of awareness campaign is required to make people familiarise with the *jowar*-based products. Health and nutritional benefits of *jowar* for all family members should be advertised exclusively. In non-traditional areas like Delhi, 'health' could be the key tool to promote the product, whereas in the areas like Hyderabad and Pune 'traditional food' could be the key point. In these traditional regions, modern and trendy product features originating from well-known traditional cereals could be the high point. The product can also be promoted in gyms and health centers as a perfect diet management product.

Place

The *kirana* (unorganised retail) is the major source of buying such products. This could be an advantageous channel with low distribution cost of product placement, but margin asked by the retailer would be very high for infant brand and newly developed product. On the other hand, placing the product in super market or at mall would attract the attention of the consumers and these places would be ideal for the campaign of the product.

Internet/online selling

Online marketing and selling is an innovative and effective way, which is gaining popularity in recent years. Many of the health labeled products are sold through this mode.

Market entry strategy

Based on the internal and external analysis of the survey report conducted under NAIP-Millets Value Chain, three types of market entry strategies are suggested. In Case 1, a client can enter all the markets with all the products, with high exposure to risks and investment. Case 2 suggests entering either all or select markets with only those products that are liked the most by the respondents. In the last case, it is suggested to be present only in the traditional sorghum-consuming markets where one can explore more options over existing sorghum products. This would be of low risk, small volume penetration in the initial stages but would eventually lead to assured demand of the product and high growth opportunity in future.

The increase in commercialisation of sorghum products can be attributed to the development and promotion of sorghum as health foods due to the initiatives taken up by the ICAR institutions such as DSR. However, there are still major challenges such as up-scaling and replication of the technologies for horizontal expansion. Further, the lower shelf-life of sorghum flour is another concern. Also, the production trends show rapid decrease in sorghum area and production during the past four decades. Therefore, in order to overcome such discrepancies, farmers are required to be directly linked with the processors and research institutions to increase their productivity and income with an objective to receive desired quality and quantity for raw material.

Market entry strategy: scenario 1 (Optimistic)

Objective: 'think global, act global'

- Corporate-level entry into the market (tie-up with FMCG Co)
- Pushing all products in all the markets
- Extensive marketing and communication campaign on Pan India level
- Utilising all marketing channels to place the product
- Targeting higher middle and affluent class
- Placing the products in major organised retail outlets

Pricing strategy

Price skimming, which entails keeping the price of the product higher to encash on initial curiosity generated through advertisements and promotional campaigns.

Market entry strategy: scenario 2 (selective-medium risk)

Objective: 'think local, act global'

- Targeting smaller and selective market share
- Ensuring collaborative entry with existing regional players
- Focusing on placing selected products in only favourable markets

Pricing strategy

Penetration pricing, which entails initially keeping the price of the product low to attract more number of consumers and maintaining that till the time consumer regularly buys the product (Table 7.6).

Table 7.6: Price strategy of *jowar* products, by city

Market	*Profile*	*Supply Option*	*Market Entry*	*Products*
Hyderabad	Traditional producer and consumer	Central manufacturing	Easy with low marketing cost	All products except *jowar* flakes
Delhi	Non-traditional but open minded	Outsourcing/ tie-up	Easy with high marketing cost	*Atta*, biscuit, and pasta
Pune	Traditional producer and consumer	Central manufacturing	Easy with medium marketing cost	*Rawa*, biscuits, and vermicelli

Market entry strategy: scenario 3 (conservative- low risk)

Objective:'think local, act local, and proceed global'

- Targeting only the traditional *jowar*-consuming areas
- Focusing more on Tier II and III cities of the traditional areas
- Marketing all the five products
- Explore other traditional areas like Rajasthan and Gujarat

Key benefits

- Least cost on advertisement and promotion of the products
- Education and awareness campaign not required
- Easy to place the products at any channel
- High volume

Pricing strategy

Cost-based pricing offers price that covers cost and has a small margin.

Suggested areas

Based on the industry secondary analysis, the study recommends that apart from the first phase target areas like Hyderabad, Pune, and New Delhi, some other states can also be targeted as consumers in these states have awareness about the consumption of *jowar* as food products. Some broad, market entry strategy and proposed effort require before entering the market is given in the Table 7.7.

Table 7.7: Potential area/markets for launching *jowar* products, by state

State	*Areas*	*Supply Options*	*Market Entry*	*Effort Required*
Andhra Pradesh	Vijaywada, Vizag, Karnool,	Central manufacturing	Easy with low marketing cost	Consumer acceptance study with launching strategy
Karnataka	Bengaluru, Dharwad, Gulberga, Bidar, Hubli	Central manufacturing	Easy with low marketing cost	Consumer acceptance study with launching strategy
Rajasthan	Jaipur, Ajmer, Jodhpur, Pali	Outsourcing/ Tie-up	Easy, moderate marketing cost	Opportunity analysis, market survey, pre-launch consumer study
Gujarat	Bharuch, Vadodara	Outsourcing/ Tie-up	Easy, moderate marketing cost	Opportunity analysis, market survey, pre-launch consumer study
Tamil Nadu	Chennai, Coimbatore	Central manufacturing	Easy with low marketing cost	Opportunity analysis, market survey, pre-launch consumer study

Impact at Micro and Macro Levels

The PPP is required in millets, especially in sorghum, wherein farmers, research institutions, entrepreneurs, and consumers are involved in value addition to this crop that has a great positive impact on yield and income of the farmers at the farm level. Here, the research institutions could develop the most suitable and high yielding seeds region-wise as per the soil and climatic conditions of the region, and by meeting the suitable stock, the entrepreneurs or processors vis-à-vis the market could rely on this for their business.

At the macro level, PPP ensures shifting of sorghum cultivation from marginal to large and more fertile land in rain-fed regions because of its higher yield and better income when compared to other crops. Sorghum thus has the potential to increase the income and standard of living at farm level. In urban areas, the consumers become used to eating healthy foods made of pure sorghum at a premium price.

Sustainability of PPP

It is a proven fact that sorghum is important to the human population meeting food and fodder demands. To promote sorghum as food product, value addition is required in each stage, which gives the optimal level of returns to consumer as well as producers/processors. Moreover, consumer-centric market economy must be supported by demand-driven farm production, procurement, storage, and handling systems. As far as the demand is concerned, 'consumption' parameter is an important determinant of millet economy. The value addition technology includes, winnowing, threshing, grading, standardising, quality control, storing, processing, packaging, transporting, branding, packing, and selling of the produce through suitable market channels. This value addition also involves the best alternative technology for the above value addition stages (units). It can identify the gaps or constraints of linkages among value-addition stages (units), helping in reducing post-harvest losses and providing suitable remedial measures.

Financial Projection: Preliminary

The preliminary financial projection is based on certain assumptions made at this time with the existing knowledge and secondary information. The business plan is intended just to demonstrate the possible returns and investments under different scenarios. One is advised to go for the preparation of a reliable detailed project report before the investment plan is finalised.

Common assumptions of different jowar products

- Target product market size is share of 1–1.5 per cent of the total demand of regular products.
- Target market share to be achieved in the span of three to five years.
- The required improvements suggested by consumers are carried out.
- Land and building cost not included.

Jowar-rich multigrain atta

Table 7.8: Capital cost requirement for *jowar*-rich multigrain *atta* processing unit

Initial investment in P&M	Rs in lakhs	7
Land and building		Nil
Profit and loss		
Sale projection	in tonnes	1000
Unit price of sale	Rs/kg	30
Revenue from sales	Rs in lakhs	300
Costs		
Raw material		225
Power/labour		15
Marketing and selling		15
Other costs (financing and depreciation)		22.45
Profitability		
Profit before tax		22.55
Tax@30%		6.765
Profit after tax		15.785
Return		5.25 %

Source: Secondary research

Based on a 20 per cent retail margin and 4 per cent VAT, the estimated retail price would be Rs 37.5/kg

Assumptions

- Raw material cost: 50 per cent *jowar* at Rs 20/kg and 50 per cent other at Rs 25/kg.
- Power and labour cost at Rs 1.5/kg; marketing and selling 5 per cent of total sales revenue.
- Other costs include interest for two months' stock at 18 per cent annually; depreciation at 10 per cent annually; and contingency 5 per cent annually.

Jowar vermicelli

Table 7.9: Capital cost requirement for *jowar* vermicelli processing unit

Initial investment in P&M	Rs in lakhs	10
Land and building		Nil
Profit and loss		
Sale projection	in tonnes	25
Unit price of sale	Rs/kg	60
Value of sale	Rs in lakhs	15
Costs		
Raw material		5
Power/labour		0.75
Marketing and selling		1.5
Other costs (financing and depreciation)		1.9
Profitability		
Profit before tax		5.85
Tax@30%		1.755
Profit after tax		4.095
Returns		27%

Source: Secondary research

Based on a 25 per cent retail margin and 4 per cent VAT, the estimated retail price would be Rs 77.50/kg.

Assumptions

- Raw material cost: 50 per cent *jowar* at Rs 20/kg and 50 per cent other at Rs 20/kg.
- Power and labour cost at Rs 3/kg; marketing and selling 10 per cent of total sales revenue
- Other costs include interest for two months' stock at 18 per cent annually; depreciation at 10 per cent annually; and contingency 5 per cent annually.

Jowar suji/rawa/samolina

Table 7.10: Capital cost requirement for *jowar suji*/*rawa*/semolina processing unit

Initial investment in P&M	Rs in lakhs	10
Land and building		Nil
Profit and loss		
Sale projection	in tonnes	300
Unit price of sale	Rs/kg	25
Revenue from sales (A) *rawa*	Rs in lakhs	75
(B) As *atta*	Rs in lakhs	75
Costs		

Contd.

Contd.

Raw material	120
Power/labour	6
Marketing and selling	7.5
Other costs (financing and depreciation)	8.35
Profitability	
Profit before tax	8.15
Tax@30%	2.445
Profit after tax	5.705
Return	7.6%

Source: Secondary research

Based on a 20 per cent retail margin and 4 per cent VAT, the estimated retail price would be Rs 31.25/kg

Assumptions

- Raw material cost: *jowar* at Rs 20/kg and recovery 50 per cent *rawa* and 50 per cent *atta.*
- Power and labour cost at Rs 2/kg; marketing and selling 10 per cent of total sales revenue.
- Other costs include interest for two months; stock at 18 per cent annually; depreciation at 10 per cent annually; and contingency 5 per cent annually.

Jowar biscuits

Table 7.11: Capital cost requirement for *jowar* biscuits processing unit

Initial investment in P&M	Rs in lakhs	30
Land and building		Nil
Profit and loss		
Sale projection	in tonnes	50
Unit price of sale	Rs/kg	90
Value of sale	Rs in lakhs	45
Costs		
Raw material		13.8
Power/labour		5
Packaging		1
Marketing and selling		13.5
Other costs (financing and depreciation)		5.66
Profitability		
Profit before tax		5.95
Tax@30%		1.78
Profit after tax		4.16
Return		9.7 %

Source: Secondary research

Based on a 30 per cent retail margin and 12.5 per cent Vat, the estimated retail price would be Rs 128/kg.

Assumptions

- Raw material cost: *jowar* at Rs 20/kg; sugar 5 per cent at Rs 35/kg; water 30 per cent at Rs 21; and additives 20 per cent at Rs 80/kg.
- Power, fuel, and labour cost at Rs 10/kg; marketing and selling 30 per cent of total sales revenue; and packaging cost Rs 3/kg.
- Other costs include interest for two months' stock at 18 per cent annually; depreciation at 10 per cent annually, and contingency 5 per cent annually.

ANNEXURE I

Sampling, Methodology, and Data Collection

As per the stated objective, the study required conducting a primary survey among the households in Hyderabad, New Delhi, and Pune. The consumers were provided with a sample kit (containing five *jowar*-based products like multigrain *atta*, *jowar* vermicelli, *jowar* flakes, *jowar suji*/*rawa*/semolina, and *jowar* biscuits) and were advised to evaluate the product on the parameters like packaging, physical appearance, and other cooking and consumption parameters like the product attributes before and after cooking, and experience encountered while preparation and consumption. The feedback of the consumers on various factors was captured through a structured questionnaire. The consumers provided their acceptability rating on the product attributes, including packaging, appearance, colour, smell, texture, preparation process, taste, etc., during both raw as well as cooked forms using 5-point hedonic scale (5 = Liked very much, 4 = Liked moderately, 3 = Neither liked nor disliked, 2 = Disliked moderately, and 1 = Disliked very much). Consumers rated prior awareness on *jowar* and its benefits and their decision to purchase or not on binary (yes/no) scale.

Sampling Plan

Sampling method

A combination of random sampling and partial judgement sampling method was adopted for the selection of the respondents because of following two criteria.

a. The targeted respondents belonged to households/families. Other segments of population such as bachelors/caterers were also covered, though at a minor level.
b. Certain segments were deliberately left out like slums, BPL population, and highly affluent class so as to avoid skewedness in the sample.

Respondents were classified into four groups/segments, based mainly on socio-economic status comprising income level, education, food habits, etc. The criteria for the sample selection included higher-income group, higher middle-income group, middle-income group, minority groups, and food habits (vegetarian/non-vegetarian).

Sample area selection

The areas within each city were identified based on the availability of the target groups of respondents in majority. Identification and selection of area were done by analysing the available secondary data, expert view, literature review based on the reports of the municipal corporation of the area, and the prevailing rent in the concerned city. The households were chosen randomly within the defined areas/locations. Details of the areas covered in all three cities are given in the Annexure.

Sample size

The sample size was determined to achieve statistically significant result at 95per cent confidence level and 5 per cent confidence interval. On this assumption, from any infinite population (above 50,000), 380 samples are required to get the desired result and accuracy level. Hence, about 400 respondents from each city were selected for the interview.

Data Collection

Questionnaire designing

The questionnaire was divided into two sections.

- Section I: Information pertaining to the demography, socio-economic conditions, and basic consumption data was collected while giving the sample to the respondents.
- Section II: Feedback on various attributes of the products like packaging, appearance in raw form, preparation process, and consumption was captured while revisiting the respondents.

Questionnaire testing

The questionnaire was tested before starting the survey to know about the likely response from the consumers and the ability of the questionnaire to meet the end results.

The questionnaire testing was done under controlled sampling technique. Around 15 samples were distributed to the consumers in Bengaluru with profile similar to that of the respondents under study.

After testing of the questionnaire, necessary modifications were incorporated to finalise it.

Field survey plan

On the basis of the sampling plan and questionnaire, the survey work was conducted in the cities of Pune, Hyderabad, and New Delhi. The interviewers were provided with sample kit, questionnaire, authorisation letter from the Directorate of Sorghum Research (DSR), and a recipe book. They were advised to follow some set guidelines while conducting interviews.

The guidelines underscored that no forceful interviews should be conducted. All the respondents should participate at their own will and there should be no coercion or monetary/award-based incentive of any kind. Once the objective of the survey was explained to the respondent, he/she should be asked whether he/she would like to participate and the interview should proceed only on affirmation. Interviews should be sourced through lawful and generally accepted ethical means. There ought to be no transgression of any legal or moral codes during the survey. The interviewer should be courteous at all times. All these guidelines were followed during the survey.

Sample kit distributed to the consumers

	Products	*Nos*	*Approximate weight*
1.	*Jowar*-rich multigrain *atta*	1	200 g
2.	*Jowar* vermicelli	1	100
3.	*Jowar* biscuits	2	25 g
4.	*Jowar* flakes/*poha*	1	150 g
5.	*Jowar suji*/*rawa*/semolina	1	100 g
6.	*Jowar* pasta (only in New Delhi)	1	100 g

ANNEXURE II

Industry Analysis and Profile of the Survey Areas

Industry Size Estimation

Multigrain atta industry

- The market size of the multigrain *atta* is very small as compared to the normal *atta* market.
- The market is estimated to be around 15,000–25,000 tonnes per annum (on an average 1–1.5 per cent of total packed *atta* industry in India).
- Mostly organised players dominate the segment and national brands are available in the market.
- Few regional players are also there in the market.
- Multigrain *atta* is sold at a premium (Ashirvaad, Pillsbury, Saffola).
- Growth of this particular segment is estimated to be 3–4 per cent annually.

Biscuit industry

- Biscuit industry size is estimated to be about 25–30 lakh tonnes per annum (Indian Biscuit's Manufacturing Association [IBMA]).
- About 55–65 per cent of the total volume catered by the organised players. Glucose biscuits constitute 45–50 per cent of the market, followed by Marie that forms 10–15 per cent of the total biscuit consumption basket.
- Share of cookies is estimated at about 3–4 per cent of the total market size.
- Growth of the overall biscuit sector has been estimated at about 12–15 per cent per annum (IBMA).

Vermicelli industry

- Indian vermicelli market size is estimated at about 30,000–40,000 tonnes per annum (*Source*: Pasta Manufacturer Association India 2009/10).
- The organised players handle 75–80 per cent of the total volume.
- Regional brands dominate in their respective regions. However, few popular regional brands like Bambino and MTR are also available in other markets.
- The sector is estimated to grow by 15 per cent annually (including pasta).

Suji/rawa/semolina industry

- Wheat and rice *suji/rawa*/semolina market size is estimated to be about 2–3 lakh tonnes per annum.
- Unorganised players dominate the sector.
- In few regions, local brands are also sold in the packed form (Shakti-Bhog, Rajdhani).
- It is also sold under private label in retail chains/stores.
- The consumption growth is estimated at about 4–5 per cent annually.

Rice flakes industry

- Rice flake market size is estimated to be about 8–10 lakh tonnes per annum. (*Source*: Foretell Internal Research).
- Unorganised players dominate the segment.
- No branded players exist in this segment.
- Rice flakes are sold under private labels in most of the urban areas.
- The consumption growth is estimated at 5–7 per cent annually.

ANNEXURE III

Consumption Pattern of Similar Products and Consumer Profile of the Survey Areas

Consumption Pattern of *Jowar* in the Study Region

Pune is the largest consumer of *jowar* among the regions under study. Snacks made from *atta* are the major forms of consumption. *Chapati* made from *jowar* flour, called *bhakari*, is the traditional food in the region. Other traditional preparations using *jowar* are *muraku, papdya, ambil,* etc.

Hyderabad is the second largest consumer of *jowar* after Pune. *Jowar* is consumed mainly as flour which is used to make *rotis*, and with the intervention of DSR various value-added products were introduced.

Delhi is a non-traditional consumer of *jowar*; migrants from traditional consuming areas and health conscious population are the main consumers of *jowar* in the city. Also to a certain extent, *jowar* is one of the major ingredients for making products like *sattu, dalia, paratha*, etc. Few companies have started selling products like *jowar*-rich multigrain *atta* and *jowar suji*/*rawa*/semolina under health food branding (as *jowar* is gluten-free, and helpful in managing weight loss and diabetis, etc.)(Fig. A1).

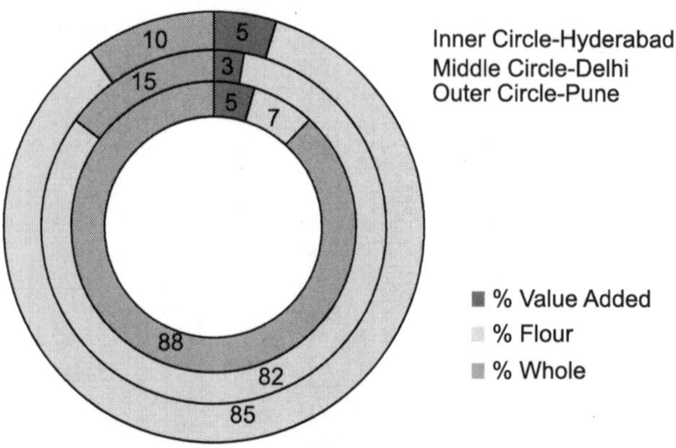

Fig. A1: Consumption pattern of *jowar* as food and in various forms (%)

Demography and Socio-economic Profile of the Country and Study Regions

In India, on an average, monthly expenditure on food items forms 40–45 per cent of the total monthly expenditure of an individual. In case of rural population, the share appears to be higher as their total expenditure is much lower than the urban population, though in absolute term size of expenditure it is much smaller as compared to the urban areas. In the urban areas, people spend more on the processed products when compared to basic food items like cereals, pulses, etc. (Table A1; Fig. A2).

Table A1: Monthly per capita expenditure (MPCE) and share of expenditure on food (%)

State	*Total Expenditure (Rs)*		*Expenditure On Food (%)*		*Percentage of Cereals on Food*	
	Rural	*Urban*	*Rural*	*Urban*	*Rural*	*Urban*
Andhra Pradesh	816	1550	52	37	15	9
Maharashtra	868	1709	48	37	13	8
Delhi	-	1847	-	37	-	6
India	772	1472	52	40	16	9

Source: NSSO (for the year 2007/08)

Socio-economic Profile of the Respondents

The respondents of the surveyed regions were classified into four groups based on their household incomes. Further to this, the profile of each class of respondents is elaborated to compare and understand their background.

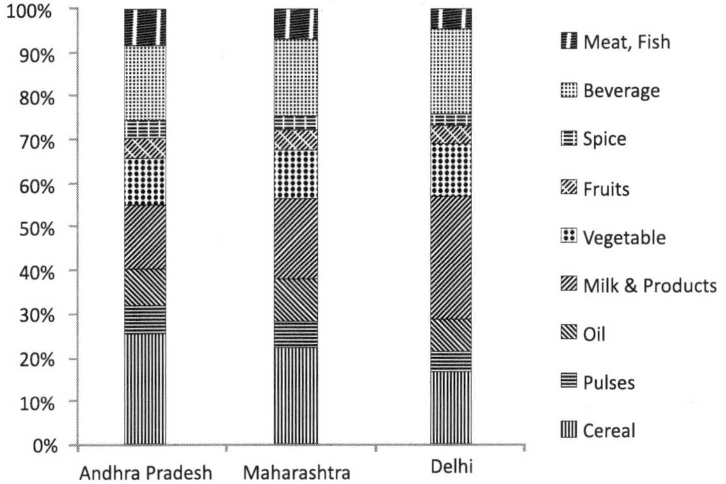

Fig. A2: Per cent expenditure on various food items in total food expenditure (urban)

The per capita expenditure of the respondents of all classes is found to be lower in Pune city and highest in the Delhi city.

Households of A 1 class have higher vegetarian ratio in all three cities. On the other hand the number of vegetarian households was high in Delhi when compared to Pune and Hyderabad (Tables A2–A4).

Table A2: Socio-economic profile of the respondents in Hyderabad city

Class	*No. of Respondents*	*Average Income*	*% Expenditure on Food*	*Per Capita Expenditure*	*Family Size*	*No. of Children*	*Food Habit (%)*
B 2	35	7528	64	1000	4	2	NV (83)
B 1	199	15339	41	1182	4	2	NV (82)
A 2	119	27500	25	1287	5	2	NV (89)
A 1	82	51034	16	1517	4	1	NV (74)

Table A3: Socio-economic profile of the respondents in Delhi city

Class	*No. of Respondents*	*Average Income*	*% Expenditure on Food*	*Per Capita Expenditure*	*Family Size*	*No. of Children*	*Food Habit (%)*
B 2	20	10000	52	1308	4	0	NV (60)
B 1	189	15740	49	1610	4	0	NV (56)
A 2	122	25676	31	1603	5	0	NV (40)
A 1	44	40965	20	1706	4	0	NV (30)

Table A4: Socio-economic profile of the respondents in Pune city

Class	*No. of Respondents*	*Average Income*	*% Expenditure on Food*	*Per Capita Expenditure*	*Family Size*	*No. of Children*	*Food Habit (%)*
B 2	45	10000	39	982	3 to 5	1	NV (73)
B 1	191	17486	28	955	4	1	NV (73)
A 2	98	27142	20	1031	4	1	NV (73)
A 1	72	41250	16	1305	4	1	NV (72)

The family composition is found to be nuclear type in Pune and Delhi, with less number of children, whereas large family size with more number of children (below 12 years) were found in Hyderabad. Average monthly household income is found to be more for all classes in Delhi, followed by Hyderabad and Pune. Majority of the female respondents are housewives, who manage their family's food budget and type of food to be purchased. Percentage share of expenses on the food for respective monthly household income groups is found to be similar in all three cities (Fig. A3).

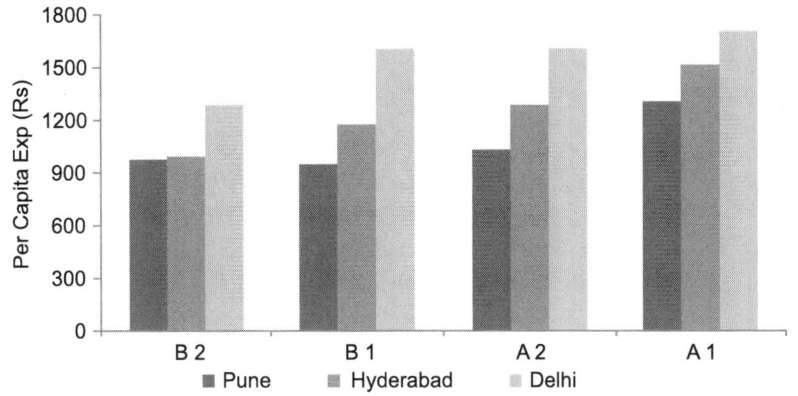

Fig. A3: Comparative per capita expenditure on food in the study region

Consumption trend of food products

Table A5: Monthly per person consumption of selected food items in India (kg)

Products	*Andhra Pradesh*		*Maharashtra*		*Delhi*		*India*	
	Rural	*Urban*	*Rural*	*Urban*	*Rural*	*Urban*	*Rural*	*Urban*
Wheat *atta*	0.144	0.455	3.272	4.000	6.6	6.386	4.192	4.350
Suji/rawa/semolina	0.110	0.228	0.073	0.099	0.037	0.032	0.047	0.098
Vermicelli/noodle	0.005	0.021	0.001	0.008	0.008	0.015	0.004	0.016
Rice flake/chira	0.033	0.014	0.088	0.110	0.010	0.005	0.064	0.062
Jowar and products	0.483	0.147	2.839	0.848	-	-	0.429	0.225
Biscuits	0.116	0.063	0.161	0.259	0.093	0.157	0.188	0.252

Source: NSSO (for the year 2004–05)

Consumption profile of various similar cereal products in the study region

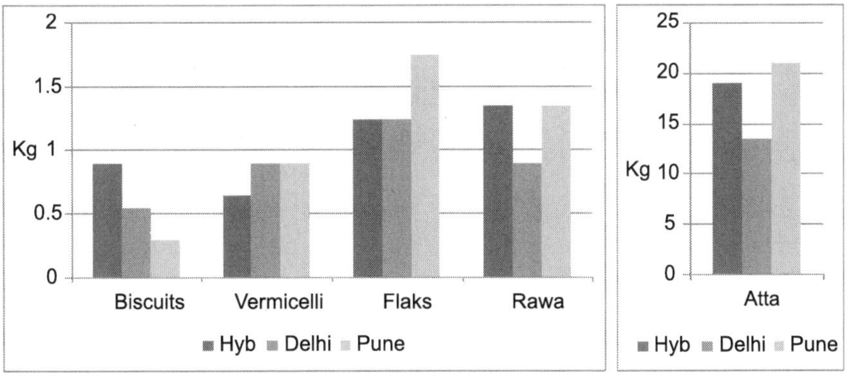

Fig. A4: City wise average monthly consumption (kg) of selected products in households

Wheat *Atta*

Atta is a very important product and has a major share in the Indian food basket. The consumers give importance to the quality of *atta* because the taste of the final product (*roti/chapati*, *poori*) is largely dependent on it. In Pune, consumers still prefer grinding wheat on their own as they perceive the packaged *atta* as being costly and of uncertain purity. On the other hand, consumers in Hyderabad and Delhi largely prefer to buy packaged *atta* as they perceive the quality of branded *atta* at par with the self ground wheat. Purchase frequency for *atta* was found to be either monthly or fortnightly and preferred SKUs are 5 kg or 10 kg for the households. In Hyderabad, majority of the surveyed consumers prefer to buy *atta* in 5 kg packs, whereas Delhi consumers prefer 10-kg pack (Table A6).

Table A6: Monthly consumption trend of wheat *atta* in households

	Pune	*Hyderabad*	*Delhi*
Quantity (kg)	18–20	12–15	20–23
Per capita (kg)	4.75	2.25	5.25
Branded: Unbranded	10:90	73:27	65:35
Name of major brand and percentage share	Self ground at *chakki*	Ashirvad (73%)	Shaktibhog (44%)
Place of major purchase	*Kirana*/Mall	*Kirana*/Retail	*Kirana*/Retail

Source: Primary Survey (*n = 400 each city*)

Wheat/Rice Vermicelli

In Pune and Delhi, consumers prefer to consume wheat vermicelli as sweet dish (made with milk, sugar, dry fruits, and flavour), whereas both sweet and salty dishes are prepared and consumed in Hyderabad. In general, two types of vermicelli, namely, roasted and normal (un-roasted), are available in the market and consumers have their own preferences based on the taste. The roasted vermicelli requires less preparation time as compared to the normal vermicelli. Purchase frequency for vermicelli is mostly monthly and preferred SKU is 200–300 g packs. In Pune and Delhi, vermicelli is consumed once or twice in a month as a sweet dish, whereas it is consumed as both salty and sweet dishes in Hyderabad more frequently (Table A7).

Table A7: Monthly consumption trend of wheat vermicelli in households

	Pune	*Hyderabad*	*Delhi*
Quantity (kg)	0.25–0.35	0.8–1.00	0.5–0.6
Per capita (kg)	0.075	0.15	0.13
Branded: Unbranded	80:20	76:24	81:19
Name of major brand and percentage share	Ganesh (50%)	Bombino (74%)	Bombino (61%)
Place of major purchase	*Kirana*/ Neighbourhood store	*Kirana*/ Neighbourhood store	*Kirana*/ Neighbourhood store

Source: Primary Survey (n = 400 each city)

Biscuits

Consumer's preference on type of biscuits depends mainly on two factors, namely, income and number of children in the family. Taste and preference of a particular type of biscuits by children influences the buying decision. Similarly, consumer's ability to spend more amounts leads to the purchase of high-priced biscuits. High-income class consumers also look at the health aspects (preferring Digestives, Lite, Oat, etc.) while buying biscuits. Cookies are mainly sold under regional brand names or in loose form. Growing awareness and health consciousness among the consumers create opportunity for biscuits claiming health benefits. Biscuits are very commonly and frequently consumed products, mainly on a daily basis. Purchase frequency varies from consumer to consumer. Our study has found the purchase of biscuits to be weekly, biweekly, and fortnightly (Table A8).

Table A8: Monthly consumption trend of biscuits in households

	Pune	*Hyderabad*	*Delhi*
Quantity (Kg)	0.8-1.0	0.5-0.75	0.75-1.0
Per capita (kg)	0.1	0.22	0.21
Branded: Unbranded	80:20	90:10	90:10
Name of major brand and percentage share	Parley-G (60 %)	Good Day (50 %) Parley-G (40 %)	Britannia & Good Day (60%)
Place of majority purchase	*Kirana*	*Kirana*	*Kirana*

Source: Primary Survey (n = 400 each city)

Rice Flake

Rice flake is also called '*poha/pohe*' in northern India and *auvalaki* in southern India. It is a very popular breakfast and snack food item, called *chivada*, in India. Maharashtra, Gujarat, Madhya Pradesh, and Rajasthan consumers mostly eat this during breakfast. Based on the thickness, three types of rice flakes are available in the market, namely, thick, medium, and thin. Consumers buy particular type of rice flake as per their requirement and preparation process. Thin size is mostly preferred for making *chivada,*whereas thick and medium sizes are used for making *poha*. Crispiness of the flake is the main attribute that the consumers prefer for making snack food, while softness is the important attribute for making *poha*. Purchasing frequency for rice flake is either fortnightly or monthly and 1 kg pack is the most common SKU among the households (Table A9).

Table A9: Monthly consumption trend of rice flakes in households

	Pune	*Hyderabad*	*Delhi*
Quantity (kg)	1.5–2.0	1.0–1.5	1.0–1.5
Per capita (kg)	0.17	0.43	0.31
Branded: Unbranded	0:100	7:93	20:80
Name of major brand and percentage share	Nil	Regional/Private label	Rajdhani (45%) and private label
Place of major purchase	*Kirana*	*Kirana*	*Kirana*

Source: Primary Survey (n = 400 each city)

Wheat *Suji*/*Rawa*/Semolina

Suji/*rawa*/semolina is an important product used for preparing breakfast as well as sweets by the Indian consumers. In southern India, it is an important ingredient for making *idli*, *dosa*, *kesari bath*, and *upma*, whereas it is consumed as sweet/*halwa* in north India. Thick/medium and fine size *rawa* is available in the market. Consumers prefer thick/medium size for making *idli*, *dosa*, and *upma* as they want granular texture and softness to remain while making the product. On the other hand, consumers prefer fine size for making sweets like *laddu*, *barfi* (cake), etc., as it gives light granular texture while swallowing (Table A10).

Table A10: Monthly consumption trend of wheat *suji*/*rawa*/semolina in households

	Pune	*Hyderabad*	*Delhi*
Quantity (kg)	1.2–1.5	1.2–1.5	0.75–1.0
Per capita (kg)	0.32	0.22	0.18
Branded: Unbranded	5:95	10:90	40:60
Name of major brand and percentage share	Private label	Regional/Private label	Rajdhani (51%) Shaktibhog (22%)
Place of major purchase	*Kirana*	*Kirana*	*Kirana*

Source: Primary Survey (n = 400 each city)

ANNEXURE IV

Survey Data Analysis and Results

The statistical analysis was conducted using the SAS statistical software program. Frequency distribution analysis was primarily done to evaluate the characteristics of the respondents and summarising their profile. Then, more than 40 per cent product-related attribute's responses were analysed under exploratory factor analyses (EFA) with varimax rotation to reduce the items to a smaller, more parsimonious set of variables. Eigen value higher than 1.0, and factor loading scores were used to identify the number of factors to extract.

Kaiser-Meyer-Olkin (KMO) measure of sampling adequacy test was done to find whether factor analysis data was appropriate within the value that falls between the ranges of being great and appropriate of factor analysis data. KMO should be 0.60 or higher in order to proceed with factor analysis. Bartlett's test was done with the result, which indicates a highly significant result with $P = 0.000$ ($P < 0.05$) and therefore factor analysis is appropriate and accepted.

Reliability of the extracted factor was assessed to examine the internal consistency of the result measurements. The reliability test was conducted using Cronbach's alpha value of 0.91 and a significance level of 0.000. The tool can be used to assess the data on Likert scale only. Due to this limitation the factor related to socio-economic and demography could not be assessed, as many of these variables were numeric.

Also, a cluster analysis exercise was carried out to group respondents into clusters based on their responses. Respondents within a cluster have a high degree of similarity, while respondents between clusters are unique. This way, cluster analysis would help us know the differentiating factors between clusters. Additionally, significant differences were identified for the cluster based on Chi-square tests and t-tests.

Logistic Regression Analysis (LRA) model was developed to analyse the important variables influencing consumer's purchasing decision to predict the correct decision made by the consumers and to compare the results among cities. The tests conducted under this model were as follows:

Steps Followed in Logistic Regression (Tests)

Multi-colinearity

It occurs when the independent variables are correlated among themselves.

Diagnosis of multi-colinearity

This is provided by the variance inflation factor (VIF). The VIF is a formal method of detecting multi-colinearity, which is widely accepted. It indicates by how much the variances of the estimated regression coefficients are inflated as compared to when the predictor variables are not correlated.

Variance inflation factor score

To understand what this means, consider the measure of the precision of a least square regression, which is given by the variance.

(VIF) = 1/(1-R 2),

Corrections: multi-colinearity

Delete the variable with the highest VIF above 2.5 one by one. Run the regression for remaining variables. Continue the procedure.

ANNEXURE V

Sample *Jowar* Product-wise Statistical Analysis Details

Jowar-rich Multigrain *Atta*

Table A11: Dimension resulted from Exploratory Factor Analysis (EFA) for *jowar* multigrain *atta*

Factor Name, Associated Variable, and Cronbach's Alpha	*Factor Load*	*Eigen Value*	*Variance Explained (%)*
Factor 1: Organoleptic (0.944)		9.191	27.85
Softness, texture, taste, smell, and overall properties of *chapati*	0.784		
Preparation (softness of dough, ease of rolling, swelling, smell)	0.735		
Appearance, texture, and colour of raw form *atta*	0.699		
Factor 2: Prior Awareness (0.926)		5.169	15.66
Jowar is a source of energy and a digestive product	0.887		
It provides wholesome nutrition to the family	0.879		
It is a perfect blend of modern and health food	0.867		
It is good for the sick and the aged	0.771		
Factor 3: Comparison with Traditional Products (0.768)		2.143	6.5
Comparison of taste with that of wheat *atta*	0.819		
Comparison of preparation process with that of wheat *atta*	0.789		
Comparison of appearance with that of wheat *atta*	0.776		
Factor 4: Socio-economic and Value for Money		1.703	7.0
Income	0.797		
Willingness to pay the price	0.531		
Expenditure on food items	-0.847		
Food habit/preferences	0.705		
Factor 5: Demography		1.340	5.0
Family size	0.826		
Number of children in the family	0.783		
Background of the family head	0.700		
Kaiser's Measure of Sampling Adequacy: (Overall MSA)			0.91906364
Bartlett's Test of Sphericity (Sig.)			0.000

A common rule of thumb for dropping the least important factors from the analysis is the K1 rule. Factors with Eigen value more than 1.0 were selected. Based on this, demography was the last factor having the value 1.340 and was selected. A Cattell Scree test plot tool was also drawn to justify the extent of component to be selected (Fig. A5).The reliability of the data and consistency in the responses received for Factor 1 pertained to 0.944, which is much higher than the measuring point of 0.600, similarly, Cronbach alpha for Factor 2 and 3 was 0.926 and 0.768, respectively. The test could not be conducted for Factor 4 and Factor 5 as these consist of independent variables and responses are in different scales. The per cent variance explained by these five factors is 62 per cent.

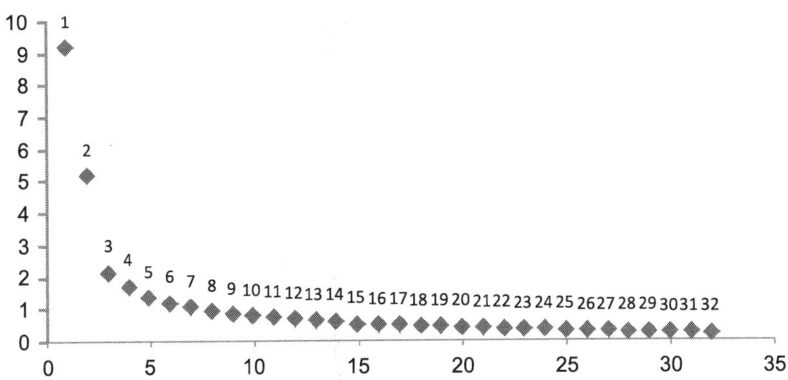

Fig. A5: Cattell Scree test plot for selecting the factor

Table A12: Profile of the important cluster identified for *jowar*-rich multigrain *atta*

Parameters		*Good Cluster*		*Average Cluster*			
		N	*%*	*N*	*%*	*Chi-Square*	*P-value*
Family size (total family members)	< 3	154	18	60	26	10.69	0.0135
	4	334	38	76	32		
	5	194	22	40	17		
	> 5	195	22	59	25		
Number of children	0	360	41	181	77	97.03	<0.0001
	1	243	28	30	13		
	2	218	25	19	8		
	> 2	57	6	5	2		

Contd.

Contd.

Parameters		*Good Cluster*		*Average Cluster*			
		N	*%*	*N*	*%*	*Chi-Square*	*P-value*
Food habit	Vegetarian	233		111		37.18	< 0.0001
	Non-vegetarian	645		124			
Occupation of the female members	Housewife	689	79	192	82	1.093	0.2957
	Working	188	21	43	18		
Occupation of the male members	Service	547					
	Other						
Monthly household income (Rs)	< 10,000	69	8	25	11	22.82	< 0.0001
	11,000–20,000	436	50	102	43		
	21,000–30,000	165	19	73	31		
	> 30,000	208	24	35	15		
Food expenditure (% income)	Up to 15	141	16	29	12	2.61	0.4557
	16–25	243	28	62	26		
	26–35	187	21	55	23		
	> 35	307	35	89	38		
Quantity consumed (kg/month)	≤ 10	454	52	65	28	45.41	< 0.0001
	11–15	74	8	38	16		
	> 15	350	40	132	56		
Buying preference	Branded	412	47	110	47	.001	0.974
	Unbranded	466	53	125	53		
Place of major purchase	*Kiranas*/Retail	620	71	181	77	23.90	< 0.0001
	Supermarket	146	17	11	5		
	Other	112	12	43	18		
Price level (Rs/kg)	< 15	121	14	96	41	154.22	< 0.0001
	15–20	330	38	120	51		
	21–25	225	26	9	4		
	26–30	194	22	10	4		
	> 30	8	1	0	0		

The result of Chi-square test and probability test on the socio-economic parameters of the clusters shows that there are significant differences (< 0.05) between both the clusters in terms of family size, children, food habit, household income, and willingness to purchase. On the other hand, the clusters do not have significant differences in terms of occupation of the female respondents and household expenditure on food out of the total income.

Attributes influencing the purchase of jowar-rich multigrain atta

In the factor analysis, we found the important dimensions from the response of the consumers with no dependent variable (purchase/not purchase). The ultimate aim is to find out the important attributes, which may or may not have an impact on the purchasing decision of the consumer.

A Logistic Regression Analysis (LRA) model was developed and step-wise analysis has been done using various statistical tools. Same techniques have been followed for the rest of the products analysed later in this chapter.

Step 1: Selection

Forward Selection Procedure was used to sort out the important variables from among a large list of variables. 'Wald' test was conducted to shortlist six to eight variables from the 43 variables collected from the respondents (Table A13).

Table A13: Shortlisted parameters after conducting Wald test

Parameter	*Estimate*	*STD Error*	*Wald Chi-square*	*Pr > Chi-square*
Intercept	-15.1888	1.5826	92.1079	< 0.0001
Stat_1	2.2029	0.318	47.9789	< 0.0001
JOAT_ERL	0.5008	0.2261	4.9055	0.0268
CHPT_CLR	0.4892	0.2393	4.1788	0.0409
CHPT_SWL	0.549	0.2285	5.7721	0.0163
CHPT_TX	0.7024	0.266	6.9729	0.0083
CHPT_TST	0.4831	0.2317	4.3484	0.037

Step 2: Correlation and multi-colinearity

These tests were done to check the correlation between the independent variables. If such a correlation was found to exist, the variables were deleted. Variance Inflation Factor (VIF) was used to further refine the variables and get the final output. Based on the VIF score, parameters/attributes having VIF of more than 2.5 were deleted and the attributes shown in Table A14 were finalised. These variables (attributes) are most important for the consumer influencing their purchase decision.

Table A14: Parameters finalised after conducting Variance Inflation Factor (VIF) test

Parameter	*t-Value*	*Variance Inflation Factor (VIF)**
Intercept	-6.8	0
Stat_1	13.74	1.09708
JOAT_ERL	4.49	1.72584
CHPT_CLR	2.41	1.90822
CHPT_SWL	3.33	1.92182
CHPT_TX	4.92	2.04676
CHPT_TST	6.18	1.96202

*Variance Inflation Score (VIS) is assumed to be ≤ 2.5. If the score was higher, then variable was rejected.

Step 3: Final modelling

After conducting a series of analysis through logistic regression techniques, attribute estimation, probability, and Odd Ratio Estimation for predicting the purchase decision for *jowar*-rich multigrain *atta* were obtained, which are represented in Table A15.

Table A15: Variables/attributes influencing the purchase decision

Factor	*MLE Estimate*	*Odds Ratio Estimate*
Intercept	-9.2045	NA
Stat_1	2.4959	12.133
JOAT_ERL	0.4605	1.585
CHPT_CLR	0.3395	1.404
CHPT_SWL	0.4801	1.616
CHPT_TX	0.6327	1.883
CHPT_TST	0.6285	1.875

Note: The analysis of maximum likelihood estimates (MLE) was used to obtain parameter estimates. Significance of parameter estimates was based on the Wald Chi-square value at $P < 0.05$.

Areas for improvement in jowar-rich multigrain atta

The *jowar*-rich multigrain *atta* has received an overwhelming response from the consumers, as it is liked and accepted by more than 90 per cent of the respondents in Hyderabad, Delhi, and Pune. The product has higher chances of success in the market from among other *jowar*-based products. Though it has various positive attributes, there are a few attributes that need to be improved for wider acceptability. Qualitative and quantitative analysis is done to find out such variables/attributes. Feedback collected from the respondents is the base for conducting qualitative analysis. Also a full variable loaded logistic regression model was built to find out the least significant variables/attributes, forming the base of quantitative method. Odds Ratio Score (ODS) estimate analysis was conducted to find out low score variables/attributes and through Maximum Likelihood Estimates (MLE), probability of each variables is calculated. Each variable/attribute was analysed and least significant or non-significant variables were rejected (low ODS variable was compared with probability to find out the least significant variable with maximum level of confidence). The same is applicable for the rest of the products analysed later in this chapter.

All 17 attributes of *jowar*-rich multigrain *atta* were used for the analysis. The attributes given in Table A16 are shortlisted as per the method of assumption in quantitative analysis. Attributes from both qualitative and quantitative analysis are presented in the next chapter.

Table A16: Attributes requiring improvement in *jowar*-rich multigrain *atta*

Parameters	*Pr > Chi-square*	*Odds Ratio*
Appearance of the *atta*	0.1573	0.701
Softness of *chapati*	0.1848	0.679
Comparison of preparation process	0.315	0.641
Comparison of taste	0.14	0.557
Overall properties of *atta*	0.367	0.749

Jowar Biscuits

Table A17: Dimensions resulted from Exploratory Factor Analysis (EFA) for *jowar* biscuits

Factor Name, Associated Variable, and Cronbach's Alpha	*Factor Load*	*Eigen Value*	*Variance Explained (%)*
Factor 1: Organoleptic (0.896)		5.91	21
Sweetness, crispiness, colour, and overall properties of biscuits	0.790		
Taste, filling, thickness	0.752		
Appearance, taste in mouth	0.743		
Smell of *jowar*	0.406		
Factor 2: Prior Awareness (0.913)		4.08	15
Jowar is a source of energy and is easy to digest	0.891		
It provides wholesome nutrition to the family	0.881		
It is a perfect blend of modern and health food	0.796		

Contd.

Contd.

Factor Name, Associated Variable, and Cronbach's Alpha	*Factor Load*	*Eigen Value*	*Variance Explained (%)*
Factor 3: Comparison with Traditional Products (0.748)		1.99	7
Comparison of overall properties with other cookies	0.760		
Smell and taste	0.697		
Crispiness and colour of the biscuits	0.610		
Factor 4: Socio-economic		1.66	6
Expenditure on food items	0.843		
Income	0.875		
Factor 5: Myth Associated with *Jowar* and Income		1.43	5
Jowar products are not suitable for all family members	0.784		
Occupation of the female members	0.436		
Factor 6: Demography		1.27	4.5
Family size	0.834		
Number of children in the family	0.752		
Kaiser's Measure of Sampling Adequacy: (Overall MSA)			0.867
Bartlett's Test of Sphericity (Sig.)			0.000

The rules explained earlier in factor analysis of *jowar*-rich multigrain *atta* were followed here as well. Demographic factor having the Eigen value of 1.27 was the last variable that was selected and justification was done by drawing Cattell Scree test plot (Fig. A6). The reliability of the data and consistency in the responses received for factor 1 pertain to 0.896, which is much higher than the measuring point of 0.600; however, the value is higher for Factor 2, which can be explained as a more reliable response. The per cent variance explained by these six factors is 58.5 per cent.

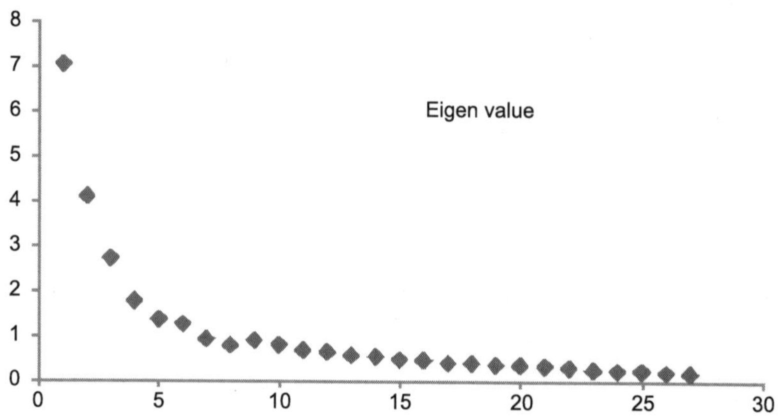

Fig. A6: Cattell Scree test plot for selecting the factor

Table A18: Profile of the important clusters identified for *jowar* biscuits

Parameters		*Good Cluster*		*Avgerage Cluster*		*Chi-square*	*P value*
		N	*%*	*N*	*%*		
Family size (total family members)	< 3	196	20	28	17	4.33	0.2279
	4	372	37	58	35		
	5	215	21	30	18		
	> 5	222	22	48	29		
Number of children	0	470	47	97	47	8.98	0.0292
	1	260	26	35	26		
	2	217	22	26	22		
	> 2	58	6	6	6		
Food habit	Vegetarian	736	73	107	65	4.47	0.0344
	Non-vegetarian	269	27	57	35		
Occupation of the female members	Housewife	209	21	33	20	.0390	0.8423
	Working	796	79	131	80		
Occupation of the male members	Service	627	63	98	60	2.07	0.3551
	Other	378	37	49	30		
Monthly household income (Rs)	< 10,000	81	8	16	10	22.84	< 0.0001
	11,000–20,000	482	48	76	46		
	21,000–30,000	205	20	55	34		
	> 30,000	237	24	17	10		
Food expenditure (% income)	Up to 15,000	165	16	13	8	18.31	0.0004
	16–25	268	27	55	34		
	26–35	206	20	50	30		
	> 35	366	36	46	28		
Quantity consumed (g/month)	≤ 100	44	4	1	1	16.66	0.0008
	110–500	309	31	74	45		
	500–100	420	42	56	34		
	> 1000	232	23	33	20		
Buying preference	Branded	928	92	130	79	28.02	< 0.0001
	Unbranded	77	8	34	21		
Place of major purchase	*Kirana*/Retail	800	80	133	81	.4094	< 0.8149
	Supermarket	141	14	20	12		
	Other	64	6	11	7		
Price level (Rs/100 g)	4–5	313	31	45	27	28.87	< 0.0001
	6–9	272	27	22	13		
	10	385	38	96	59		
	11–15	35	3	1	1		

Attributes influencing the purchase of jowar biscuits

Step 1: Sorted attributes after forward selection process (MLE)

Table A19: Shortlisted attributes after Forward Selection Process (MLE)

Parameter	*Estimate*	*STD Error*	*Wald Chi-square*	*Pr > Chi-square*
Intercept	-6.1844	0.8129	57.8856	< 0.0001
% Food_Exp	0.0209	0.0097	4.6313	0.0314
JOBS_MFL	0.679	0.2169	9.8023	0.0017
JOBS_SWT	1.0626	0.2158	24.2439	< 0.0001
Com_TST	1.2507	0.3565	12.3116	0.0005
Com_OAL	1.1927	0.3672	10.5533	0.0012

Step 2: Correlation and multi-colinearity

Table A20: Parameters finalised after conducting Variance Inflation Factor test

Parameter	*t- Value*	*Variance Inflation Factor (VIF)**
Intercept	1.72	0
JOBS_SWT	6.78	1.74414
JOBS_MFL	4.97	1.55222
Com_SML	6.34	1.0706
JOBS_CRS	4.89	1.57991

*Variance Inflation Score (VIS) is assumed to be ≤ 2.5. If the score is high, then the variable is rejected.

Step 3: Final modelling

Table A21: Factors shortlisted after conducting Maximum Likelihood and Odd Ratio Estimates

Factor	*MLE Estimate*	*Odds Ratio Estimate*
Intercept	-6.2408	NA
JOBS_SWT	0.9833	2.673
JOBS_MFL	0.5173	1.677
Com_SML	1.4114	4.102
JOBS_CRS	0.5185	1.68

Note: The analysis of Maximum Likelihood Estimates (MLE) was used to obtain parameter estimates. Significance of parameter estimates was based on the Wald Chi-square value at $P < 0.05$

Areas for improvement in attributes of jowar biscuits

Here, Odds Ratio Score (ODS) estimate analysis was conducted to find out less significant or non-significant variables/attributes of *jowar* biscuits. After qualitative analysis, these attributes were recommended for improvement (Table A22).

Table A22: Attributes of *jowar* biscuits requiring improvements

Attribute	*Pr > Chi-square*	*Odds Ratio*
JOBS appearance	0.823	0.5378
JOBS colour	0.828	0.5087
JOBS taste	0.82	0.4858
JOBS smell	0.736	0.3369
Crispiness compared to traditional	0.482	0.1057
Colour compared to traditional	0.869	0.7589

Jowar Vermicelli

Table A23: Dimensions resulted from Exploratory Factor Analysis (EFA) for *jowar* vermicelli

Factor Name, Associated Variable, and Cronbach's Alpha	*Factor Load*	*Eigen Value*	*Variance Explained (%)*
Factor 1: Organoleptic (0.950)		8.82	28
Overall experience with cooked and raw vermicelli	0.837		
Appearance, colour, texture, and stickiness of cooked vermicelli	0.776		
Colour of raw vermicelli and smell of cooked vermicelli	0.766		
Preparation time, thickness, and appearance of raw vermicelli	0.745		
Factor 2: Prior Awareness (0.914)		4.02	13
Jowar is a source of energy and is a digestive product	0.881		
It provides wholesome nutrition to the family	0.867		
It is a perfect blend of modern and health food	0.818		
Factor 3: Comparison with Traditional Products (0.699)		1.81	6
Comparison of appearance with that of wheat vermicelli	0.752		
Comparison of taste with that of wheat vermicelli	0.714		
Comparison of the preparation process with that of wheat Vermicelli	0.657		

Contd.

Contd.

Factor Name, Associated Variable, and Cronbach's Alpha	*Factor Load*	*Eigen Value*	*Variance Explained (%)*
Factor 4: Socio-economic and Value for Money		1.65	5
Income	0.876		
Expenditure on food items	0.823		
Willingness to purchase	-0.360		
Factor 5: Myth associated with *Jowar* and Income		1.36	4.5
Jowar products are not suitable for all family members	0.784		
Occupation of the female member	0.446		
Factor 6: Demography		1.28	4.0
Family size	0.848		
Number of children in the family	0.736		
Background of the family head	0.232		
Kaiser's Measure of Sampling Adequacy: (Overall MSA)			0.917
Bartlett's Test of Sphericity (Sig.)			0.000

The reliability of the data and consistency in the responses received for factor 1 pertain to 0.917, which is much higher than the measuring point of 0.600. Cronbach alpha for Factors 1, 2, and 3 is 0.950, 0.914, and 0.699, respectively. The per cent variance explained by these six factors is 59.5 per cent (Table A23; Fig. A7).

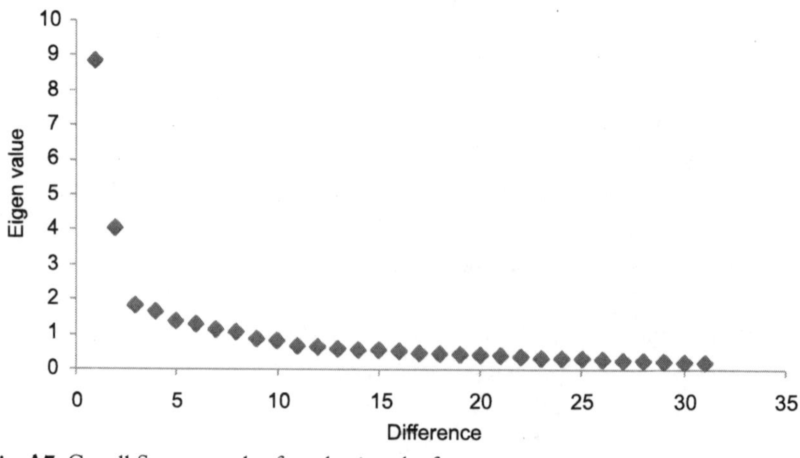

Fig. A7: Cattell Scree test plot for selecting the factor

Table A24: Profile of the important cluster identified for *jowar* vermicelli

Parameters		*Good Cluster*		*Average. Cluster*		*Chi-square*	*P value*
		N	*%*	*N*	*%*		
Family size (total family members)	< 3	108	18	112	20	2.94	.4016
	4	229	39	199	35		
	5	121	21	122	21		
	> 5	126	22	141	25		
Number of children	0	233	40	328	57	42.52	< .0001
	1	159	27	128	22		
	2	144	25	102	18		
	> 2	48	8	16	3		
Food habit	Non-vegetarian	439	75	392	68	6.76	.009
	Vegetarian	145	25	182	32		
Occupation of the female members	Housewife	490	84	426	74	16.44	< .0001
	Working	94	16	148	26		
Occupation of the male members	Service	360	61	355	62	.0194	.990
	Other	224	39	219	38		
Monthly household income (Rs)	< 10,000	35	6	60	10	23.57	< .0001
	11,000-20,000	272	47	284	49		
	21,000-30,000	120	21	136	24		
	> 30,000	157	27	94	16		
Food expenditure (% Income)	Up to 15	99	17	74	13	16.23	.001
	16-25	158	27	165	29		
	26-35	101	17	148	26		
	> 35	226	39	187	33		
Quantity consumed (g/month)	≤100	24	4	8	1	23.57	< .0001
	110-500	304	52	357	62		
	500-100	199	34	140	24		
	> 1000	57	10	69	12		
Buying preference	Branded	440	75	407	71	2.90	.088
	Unbranded	144	25	167	29		
Place of major purchase	*Kirana*/Retail	450	77.0	411	71.60	21.56	< .8149
	Supermarket	116	19.8	107	18.64		
	Other	18	3.08	56	9.75		
Price level (Rs/200 g)	5-9	118	20	100	17	18.45	.001
	10	189	32	252	44		
	11-15	173	30	150	26		
	16-20	77	13	57	10		
	21-35	27	5	15	3		

Attributes influencing the purchase of jowar vermicelli

Step 1: Sorted attributes after forward selection process (MLE)

Table A25: Sorted attributes after Forward Selection Process (MLE)

Parameter	*Estimate*	*STD Error*	*Wald Chi-square*	*Pr > Chi-square*
Intercept	-7.526	0.8549	77.5051	< 0.0001
Stat_6	0.9218	0.247	13.9272	0.0002
JOVM_SML	0.8258	0.1765	21.8967	< 0.0001
VMSW_STK	0.389	0.1605	5.8735	0.0154
VMSW_OAL	0.5341	0.1924	7.7053	0.0055
Price	-0.0667	0.0227	8.6837	0.0032
Com_APR	0.9273	0.2726	11.5666	0.0007
Com_PRP	1.1406	0.2615	19.0207	< 0.0001
Com_TST	1.8336	0.2502	53.7283	< 0.0001

Step 2: Correlation and multi-colinearity

Table A26: Parameters shortlisted after Variance Inflation Score estimate

Parameter	*t- Value*	*Variance Inflation Score (VIS)**
Intercept	-6.29	0
Stat_6	4.24	1.04897
JOVM_SML	6.35	1.8001
VMSW_STK	3.36	1.86571
VMSW_OAL	4.09	2.1505
Com_APR	3.68	1.42049
Com_PRP	6.16	1.43476
Com_TST	8.96	1.38208

* Variance Inflation Score (VIS) is assumed to be ≤ 2.5. If the score is high, then the variable is rejected.

Step 3: Final modelling

Table A27: Parameters finalised after Maximum Likelihood and Odds Ratio Estimates

Factor	*MLE Estimate*	*Odds Ratio Estimate*
Intercept	0.8243	NA
Smell of *jowar* vermicelli	0.8181	2.266
Stickiness of sweet vermicelli	0.3982	1.489
Overall qualities of sweet vermicelli	0.5142	1.672
Com_APR	0.8984	2.456
Com_PRP	1.1763	3.242
Com_TST	1.8347	6.263

Note: The analysis of Maximum Likelihood Estimates (MLE) was used to obtain parameter estimates. Significance of parameter estimates was based on the Wald Chi-square value at $P < 0.05$.

Areas for improvement in attributes of jowar vermicelli

Table A28: Attributes requiring improvement in *jowar* vermicelli

Attribute	*Pr > Chi-square*	*Odds Ratio*
JOVM_Appearance	0.866	0.5151
JOVM_Thickness	0.958	0.843
JOVM_Colour	0.787	0.2847
Cooked Vermi_Colour	0.877	0.5368

Along with the result of above quantitative analysis, consumer's feedback and qualitative observations were included and used as a basis for improvement in the product.

Jowar *Suji/Rawa*/Semolina

Uppit/Upma

Table A29: Dimensions resulted from Exploratory Factor Analysis (EFA) for *jowar suji/rawa*/semolina (*upma/uppit*)

Factor Name, Associated Variable, and Cronbach's Alpha	*Factor Load*	*Eigen Value*	*Variance Explained (%)*
Factor 1: Organoleptic (0.946)		9.38	29
Taste, overall qualities, and softness of *uppit*	0.810		
Appearance, water absorption, and texture of *uppit*	0.750		
Colour and *rawa* and *uppit*	0.736		
Smell of *rawa* and *uppit*	0.716		
Texture of *rawa* and ease of roasting	0.702		
Factor 2: Prior Awareness (0.720)		2.58	8
Jowar product is a high energy source and easy to digest	0.781		
It is perfect blend of of modern and health food	0.661		
It provides wholesome nutrition to the family	0.371		
Factor 3: Comparison with Traditional Products (0.697)		2.45	7.5
Comparison of taste with that of wheat *rawa*	0.688		
Comparison of preparation process with that of wheat *rawa*	0.667		
Comparison of appearance with that of wheat *rawa*	0.651		
Factor 4: Socio-economic and Value for Money		1.57	5
Income	0.810		
Price of the product	0.621		
Expenditure on food items	-0.771		

Contd.

Contd.

Factor Name, Associated Variable, and Cronbach's Alpha	*Factor Load*	*Eigen Value*	*Variance Explained (%)*
Factor 5: Myth Associated with *Jowar* and Income		1.50	4.75
Jowar products are not suitable for all family members	0.759		
Occupation of the female members	0.453		
Factor 6: Demography		1.40	4.0
Family size	0.810		
Number of children in the family	0.743		
Kaiser's Measure of Sampling Adequacy: (Overall MSA)			0.892
Bartlett's Test of Sphericity (Sig.)			.000

Suji/*rawa*/semolina is used for making different types of food items in Pune, Delhi, and Hyderabad. In Hyderabad, respondents make *upma*/*uppit* from *rawa*, whereas Delhi respondents prepare *sira*/*halwa*/Sweet. In Pune, respondents prepare both *sira* and *upma*. Both types of food have different attributes in terms of preparation and taste. To map correctly the attributes of the respondents, the product has been divided into two groups and analysed for better clarity and results (Table A29).

Cattell Scree test plot was drawn to select the extent of component (Fig. A8) and based on this, demographic as the last factor having a value of 1.40 was selected. The reliability of the data and consistency in the responses received for factor 1 pertain to 0.946, which is much higher than the measuring point of 0.600; the value is lower for Factors 2 and 3, which can be explained comparatively less important response. The per cent variance explained by these six factors is 56.25 per cent.

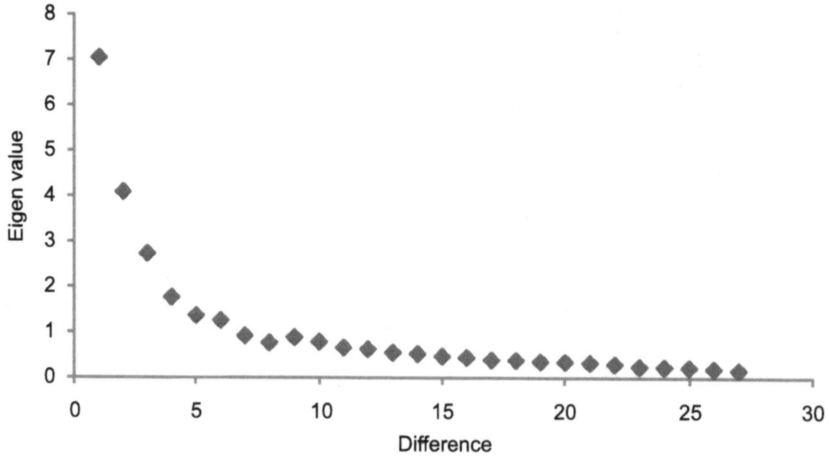

Fig. A8: Cattell Scree test plot for selecting the factor

Table A30: Profile of the important cluster identified for *jowar rawa*

Parameters		*Good Cluster*		*Avgerage Cluster*		*Chi-square*	*P value*
		N	*%*	*N*	*%*		
Family size (total family members)	< 3	74	19	54	19	1.86	0.600
	4	155	40	105	38		
	5	73	19	64	23		
	> 5	84	22	55	20		
Number of children	0	125	32	125	45	21.45	< 0.0001
	1	109	28	83	30		
	2	114	30	62	22		
	> 2	38	10	8	3		
Food habit	Vegetarian	303	78	216	78	0.060	0.8057
	Non-vegetarian	83	22	62	22		
Occupation of the female member	Housewife	73	19	79	28	8.27	0.0080
	Working	313	81	199	72		
Occupation of the male member	Service	231	60	181	65	4.69	0.096
	Other	155	40	97	35		
Monthly household income (Rs)	< 10,000	24	6	31	11	32.06	< 0.0001
	11,000-20,000	150	39	150	54		
	21,000-30,000	72	19	47	17		
	> 30,000	140	36	50	18		
Food expenditure (% income)	Up to 15	95	25	42	15	16.78	0.0008
	16-25	124	32	73	26		
	26-35	68	18	63	23		
	> 35	99	26	100	36		
Quantity consumed (g/month)	≤ 100	104	27	73	26	1.14	0.5658
	110-500	185	48	125	45		
	500-1000	97	25	80	29		
Buying preference	Branded	12	3	18	6	4.62	0.031
	Unbranded	370	96	249	90		
Place of major purchase	*Kirana*/Retail	289	75	202	73	17.15	0.0002
	Supermarket	81	21	42	15		
	Other	16	4	34	12		
Price level (Rs/100 g)	4-5	180	47	176	63	18.52	0.0003
	6-9	104	27	53	19		
	10	83	22	42	15		
	11-15	19	5	7	3		

Attributes influencing the purchase of jowar rawa (uppit)

Step 1: Sorted attributes after forward selection process (MLE)

Table A31: Attributes after Forward Selection Process (MLE)

Parameter	*Estimate*	*STD Error*	*Wald Chi.Sq.*	*Pr > Chi Sq.*
Intercept	-4.9287	0.9513	26.8439	< .0001
Stat_2	-0.8521	0.3328	6.5552	0.0105
JOSJ_A	-0.7027	0.2305	9.2947	0.0023
UPIT_A	0.6162	0.2545	5.8611	0.0155
UPIT_SML	0.6172	0.2618	5.5582	0.0184
UPIT_TX	0.7608	0.2359	10.4	0.0013
UPIT_TST	0.3059	0.32	0.9138	0.3391
Com_PRP	1.1375	0.4068	7.8182	0.0052
Com_TST	0.9937	0.3952	6.3237	0.0119

Step 2: Correlation and multi-colinearity

Table A32: Parameters after Variance Inflation Factor analysis

Parameter	*t- Value*	*Variance Inflation Factor (VIF)**
Intercept	-2.92	0
UPIT_A	3.73	2.10511
UPIT_SML	1.74	1.81641
UPIT_TST	4.09	3.13943
Com_PRP	6.17	1.77151
Com_TST	4.43	1.72839

* Variance Inflation Score (VIS) is assumed to be ≤ 2.5. If the score is high, then the variable is rejected.

Step 3: Final modelling

Table A33: Parameters finalised after Maximum Likelihood and Odd Ratio Estimates

Factor	*MLE Estimate*	*Odds Ratio Estimate*
Appearance of *uppit*	0.6117	1.844
Smell of *uppit*	0.4439	1.559
Taste of *uppit*	0.6917	1.997
Comparison of the preparation process with that of wheat *rawa*	1.3283	3.775
Comparison of taste with that of wheat *rawa*	0.9826	2.671

Note: The analysis of maximum likelihood estimates (MLE) was used to obtain parameter estimates. Significance of parameter estimates was based on the Wald Chi-square value at $P < 0.05$.

Areas for improvement in attributes of jowar rawa (uppit)

Table A34: Attributes requiring improvement in *jowar rawa* (*uppit*)

Attribute	*Pr > Chi-square*	*Odds Ratio*
JOSJ_A	0.61	0.1603
JOSJ_TX	0.859	0.6421
JOSJ_ROS	0.782	0.4888
JOSJ_OAL	0.605	0.254
UPIT_SFT	0.933	0.8325
UPIT_OAL	0.854	0.721

Sira/Sweet/*Halwa*

Table A35: Dimensions resulted from Exploratory Factor Analysis (EFA) for *jowar rawa* (*sira*)

Factor Name, Associated Variable, and Cronbach's Alpha	*Factor Load*	*Eigen Value*	*Variance Explained (%)*
Factor 1: Organoleptic (0.940)		8.13	25
Overall qualities and appearance of *sira*	0.808		
Texture and water absorption of *sira*	0.767		
Smell, colour, and softness of *sira*	0.755		
Overall qualities in raw form, ease of roasting, colour, and taste of *sira*	0.736		
Texture, appearance, and smell of *rawa*/*suji*	0.690		
Factor 2: Prior Awareness (0.917)		4.91	15
Jowar is a source of high energy and easy to digest	0.880		
It is perfect blend of modern and health food	0.860		
It provides wholesome nutrition to the family	0.854		
Jowar products are not suitable for all family members	0.673		
Factlor 3: Comparison with Traditional Products (0.641)		1.68	5
Comparison of appearance with that of wheat *rawa*	0.753		
Comparison of taste with that of wheat *rawa*	0.726		
Comparison of preparation process with that of wheat *rawa*	0.720		
Factor 4: Socio-economic		1.64	5
Income	0.800		
Occupation of the female members	-0.510		
Expenditure on food items	-0.713		

Contd.

Contd.

Factor Name, Associated Variable, and Cronbach's Alpha	*Factor Load*	*Eigen Value*	*Variance Explained (%)*
Factor 5: Demography		1.38	4.0
Family size	0.810		
Number of children in the family	0.743		
Factor 6: Value for Money and Food Habits		1.12	3.5
Price of the product	0.663		
Vegetarian/non-vegetarian	0.361		
Kaiser's Measure of Sampling Adequacy: (Overall MSA)			0.901
Bartlett's Test of Sphericity (Sig.)			.000

As per the K1 rule, the factors with Eigen value more than 1.0 are selected. Here, value for money and food habit combination was the last factor having a value 1.12. A Cattell Scree test plot tools was also drawn to justify the extent of component to be selected (Fig. A9). The reliability of the data and consistency in the responses received for Factor 1 pertain to 0.940, which is much higher than the measuring point of 0.600, also higher from Factors 2 and 3, which can be explained as more reliable response. The per cent variance explained by these six factors is 57.5 per cent.

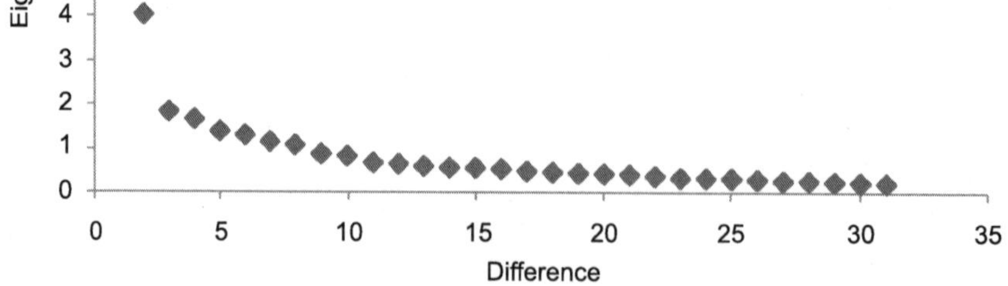

Fig. A9: Cattell Scree test plot for selecting the factor

Table A36: Profile of the important clusters identified for *jowar rawa*

Parameters		*Good Cluster*		*Avgerage Cluster*		*Chi-square*	*P value*
		N	*%*	*N*	*%*		
Family size (total family members)	< 3	28	25	26	13	6.29	0.098
	4	42	37	79	41		
	5	22	19	46	24		
	> 5	22	19	43	22		
Number of children	0	95	83	109	56	24.11	< 0.0001
	1	11	10	46	24		
	2	8	7	36	19		
	> 2	0	0	3	2		
Food habit	Vegetarian	55	48	109	56	1.82	0.1775
	Non-vegetarian	59	52	85	44		
Occupation of the female members	Housewife	15	13	32	16	.6183	0.4317
	Working	99	87	162	84		
Occupation of the male members	Service	64	56	125	64	3.53	0.1710
	Other	50	44	69	36		
Monthly household income (Rs)	< 10,000	13	11	14	7	3.92	0.2706
	11,000–20,000	61	54	118	61		
	21,000–30,000	25	22	46	24		
	> 30,000	15	13	16	8		
Food expenditure (% Income)	Up to 15	8	7	10	5	1.61	0.6591
	16–25	21	18	46	24		
	26–35	25	22	44	23		
	> 35	60	53	94	48		
Quantity consumed (g/month)	101–500	81	71	79	41	27.92	< 0.0001
	501–1000	25	22	72	37		
	> 1000	8	7	43	22		
Buying preference	Branded	46	40	53	27	5.59	0.0081
	Unbranded	68	60	141	73		
Place of major purchase	*Kirana*/Retail	109	96	166	86	14.29	0.0008
	Supermarket	4	4	3	2		
	Other	1	1	25	13		
Price level (Rs/100 g)	10–19	53	46	98	51	2.75	0.4310
	20–25	32	28	58	30		
	26–30	24	21	27	14		
	31–45	5	4	11	6		

Attributes influencing purchase of jowar rawa (sira)

Step 1: Sorted attributes after forward selection process (MLE)

Table A37: Finalised attributes after Forward Selection Process (MLE)

Parameter	*Estimate*	*STD Error*	*Wald Chi-square*	*Pr > Chi-square*
Intercept	-7.8423	1.3737	32.5917	< 0.0001
Stat_5	0.7681	0.2889	7.0702	0.0078
JOSJ_TX	0.4503	0.2278	3.907	0.0481
JOSJ_CLR	0.7214	0.2232	10.4451	0.0012
SIRA_SML	0.5797	0.1878	9.5286	0.002
price	-0.0601	0.0252	5.6737	0.0172
Com_APR	1.4304	0.2949	23.5282	< .0001
Com_PRP	1.1809	0.3176	13.8295	0.0002
Com_TST	1.5624	0.3293	22.5168	< .0001

Step 2: Correlation and multi-colinearity

Table A38: Final attributes after Variance Inflation Factor analysis

Parameter	*t- Value*	*Variance Inflation Factor (VIF)**
Intercept	-6.13	0
Stat_5	3.52	1.02231
JOSJ_TX	1.72	1.4832
JOSJ_CLR	3.41	1.59764
SIRA_SML	3.26	1.46368
Com_APR	5.03	1.26496
Com_PRP	4.13	1.27288
Com_TST	4.97	1.28722

* Variance Inflation Score (VIS) is assumed to be ≤ 2.5.If the score is high, then the variable is rejected.

Step 3: Final modelling

Table A39: Final parameters after Maximum Likelihood and Odds Ratio Estimates

Factor	*MLE Estimate*	*Odds Ratio Estimate*
JOSJ_TX	0.3834	1.467
JOSJ_CLR	0.8098	2.247
SIRA_SML	0.5719	1.772
Com_APR	1.4385	4.214
Com_PRP	1.1588	3.186
Com_TST	1.5201	4.573

Note: The analysis of Maximum Likelihood Estimates (MLE) was used to obtain parameter estimates. Significance of parameter estimates was based on the Wald Chi-square value at P < 0.05.

Areas for improvement in attributes of jowar rawa (sira)

Table A40: Attributes requiring improvement in *jowar rawa* (*sira*)

Attribute	*Pr > Chi-square*	*Odds Ratio*
Overall attributes of *rawa*	0.216	0.723
Appearance of *sira*	0.4365	0.809
Colour of *sira*	0.5908	0.868
Water absorption (*sira*)	0.8203	0.944

Jowar Flakes

Table A41: Dimensions resulted from Exploratory Factor Analysis (EFA) for *jowar* flakes

Factor Name, Associated Variable, and Cronbach's Alpha	*Factor Load*	*Eigen Value*	*Variance Explained (%)*
Factor 1: Organoleptic (0.952)		8.44	28
Overall experience with raw and cooked (*poha*) flakes	0.830		
Water absorption, softness, colour, and smell of cooked flakes (*poha*)	0.810		
Taste and stickiness of *poha*, and colour and smell of the *jowar* flakes	0.789		
Thickness and appearance of *jowar* flakes and appearance of *poha*	0.757		
Factor 2: Prior Awareness (0.914)		4.20	14
Jowar is a source of energy and is a digestive	0.860		
It provides wholesome nutrition to the family	0.876		
It is a perfect blend of modern and health food	0.814		
Factor 3: Myth Associated with *Jowar* (0.732)		1.89	6
Jowar products are not suitable for all family members	0.804		
Factor 4: Comparison with Traditional Products (0.572)		1.53	5
Comparison of appearance with that of wheat vermicelli	0.801		
Comparison of preparation process with that of wheat vermicelli	0.752		
Comparison of taste with that of wheat vermicelli	0.566		

Contd.

Contd.

Factor Name, Associated Variable, and Cronbach's Alpha	*Factor Load*	*Eigen Value*	*Variance Explained (%)*
Factor 5: Socio-economic and Value for Money		1.40	4.5
Income	0.879		
Expenditure on food items	-0.829		
Occupation of the female member	-0.266		
Factor 6: Demography		1.27	4.0
Family size	0.843		
Number of children in the family	0.763		
Food habit	-0.265		
Kaiser's Measure of Sampling Adequacy: (Overall MSA)			0.917
Bartlett's Test of Sphericity (Sig.)			0.000

The reliability of the data and consistency in the responses received for Factor 1 pertain to 0.952, which is much higher than the measuring point of 0.600. Cronbach alpha for Factors 2, 3, and 4 is 0.914, 0.732, and 0.572, respectively. The per cent variance explained by these six factors is 61.5 per cent (Table A41; Fig. A10).

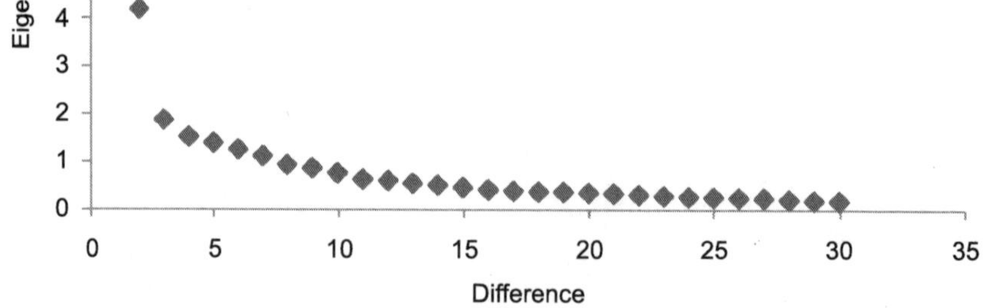

Fig. A10: Cattell Scree test plot for selecting the factor

Table A42: Profile of the important clusters identified for *jowar* flakes

Parameters		*Good Cluster*		*Average Cluster*		*Chi-square*	*P value*
		N	*%*	*N*	*%*		
Family size (total family members)	< 3	124	18	51	22	7.37	0.0610
	4	269	39	70	30		
	5	151	22	47	20		
	> 5	149	22	62	27		
Number of children	0	288	42	173	75	82.44	< .0001
	1	186	27	37	16		
	2	171	25	16	7		
	> 2	48	7	4	2		
Food habit	Non-vegetarian	530	76	117	51	54.03	< .0001
	Vegetarian	163	24	113	49		
Occupation of the female member	Housewife	531	77	191	83	4.18	0.0408
	Working	162	23	39	17		
Occupation of the male member	Service	431	62	128	56	5.76	0.0561
	Other	262	38	102	44		
Monthly household income (Rs)	< 10,000	54	8	21	9	29.51	< .0001
	11,000–20,000	313	45	103	45		
	21,000–30,000	138	20	77	33		
	> 30,000	188	27	29	13		
Food expenditure (% income)	Up to 15	129	19	22	10	12.62	0.0055
	16–25	174	25	61	27		
	26–35	126	18	57	25		
	> 35	264	38	90	39		
Quantity consumed (g/month)	≤ 100	213	31	91	40	12.28	.0065
	110–500	276	40	63	27		
	500–100	163	24	59	26		
	> 1000	41	6	17	7		
Buying preference	Branded	43	6	41	18	28.19	< .0001
	Unbranded	650	94	189	82		
Place of major purchase	*Kirana*/Retail	538	78	208	90	18.26	.0001
	Supermarket	114	16	16	7		
	Other	41	6	6	3		
Price level (Rs/1 kg)	8–10	14	2	0	0	63.71	< .0001
	11–15	115	17	7	3		
	16–20	274	40	139	60		
	21–25	147	21	63	27		
	26–30	119	17	21	9		
	31–45	24	3	0	0		

Attributes influencing the purchase of jowar flakes

Step 1: Sorted attributes after forward selection process (MLE)

Table A43: Final attributes after Forward Selection Process (MLE)

Parameter	*Estimate*	*STD Error*	*Wald Chi-square*	*Pr > Chi Sq.*
Intercept	-6.9078	0.6142	126.4815	< .0001
V_NV	-0.4943	0.1899	6.7737	0.0093
Food_Exp	-0.0117	0.00479	5.9367	0.0148
Stat_2	-0.7311	0.1836	15.8617	< .0001
Stat_7	1.1588	0.1909	36.8631	< .0001
POHA_A	0.6428	0.1524	17.8023	< .0001
POHA_WAB	0.601	0.133	20.4298	< .0001
POHA_SFT	0.4272	0.1421	9.0353	0.0026
POHA_OAL	0.4376	0.1543	8.0442	0.0046
Com_PRP	0.819	0.239	11.7413	0.0006

Step 2: Correlation and multi-colinearity

Table A44: Final attributes after Variation Inflation Factor analysis

Parameter	*t- Value*	*Variance Inflation Factor (VIF)**
Intercept	-8.16	0
V_NV	2.04	1.02636
Stat_7	5.34	1.03395
POHA_A	4.16	2.05193
POHA_WAB	4.74	2.31312
POHA_SFT	4.17	2.30652
POHA_OAL	3.86	2.52836
Com_PRP	2.3	1.07143

* Variance Inflation Score (VIS) is assumed to be ≤ 2.5. If the score is high, then the variable is rejected.

Step 3: Final modelling

Table A45: Shortlisted attributes after Maximum Likelihood and Odds Ratio Estimates

Factor	*MLE Estimate*	*Odds Ratio Estimate*
Stat_7	1.0094	2.744
POHA_A	0.5612	1.753
POHA_WAB	0.5829	1.791
POHA_SFT	0.458	1.581
POHA_OAL	0.49	1.632
Com_PRP	0.7902	2.204

Note: The analysis of Maximum Likelihood Estimates (MLE) was used to obtain parameter estimates. Significance of parameter estimates was based on the Wald Chi-square value at $P < 0.05$.

Areas for improvement in attributes of jowar flakes

Table A46: Attributes requiring for improvement in *jowar* flakes

Attribute	*Pr > Chi-square*	*Odds Ratio*
Appearance of *jowar* flakes	0.8782	0.972
Smell of *jowar* in *poha*	0.6893	0.94
Comparison of appearance with that of rice flakes	0.0632	0.619

ANNEXURE VI

SWOT and Competitive Analysis

SWOT Analysis of *Jowar* Food Products

Strengths

- There is a prior awareness on *jowar* being a healthy and nutritious food.
- Good response for quality certification from NIN.
- Strong research and development facility is available.
- DSR's brand 'Eatrite' is the only brand in the market offering versatile product range based on *jowar*.
- Easy to be adopted in terms of appearance, preparation, and taste when compared with traditional products.
- Taste of the product was liked by the respondents.
- People are willing to consider the product if 'Eatrite' is launched in the market.

Weaknesses

- New brand in the market.
- Lack of awareness about the product among the non-traditional respondents.
- Lack of commercial business operation and good sales/distribution network; and there is limited presence in the market.
- Immature economy of scale.
- Lack of promotion of the product.
- Shelf-life of the product is short (one month).
- The typical smell of *jowar* is disliked by the non-traditional consumers.
- Difficulties are encountered in the preparation/cooking of *jowar* by urban consumers.
- Appearance of the *jowar* is inferior as compared with the traditional product.

Opportunity

- Growing willingness in consumers to try new products with health benefits.
- Diabetic and obese consumers are happy with the product having major health benefits for them.
- A good substitute for traditional products in terms of variety.
- Health and Wellness food market is of the size Rs 10,000 crore and expected to grow by CAGR 23 per cent (TSMG).
- Increasing disposable incomes, diminishing culinary skills, and the rising need for convenience food would be the driving factors for making *jowar* products popular in traditional areas.

Threat

- Initial hesitation by the retailers to market the products in non-traditional areas.
- Consistency about the quality is doubtful in the consumer's mind, as the supplier is not a branded fast moving consumer goods (FMCG) supplier.
- Sustainability in the short term is high due to marketing set-up cost.
- Increasing raw material price and uncertainly in the production of *jowar*.
- Negative perceptions with regard to the health (non-edible varieties in some seasons; issues of dental disease).
- The health product segment size is very small and many consumers fall in the 'Late Majority' category.

Product Wise SWOT Analysis

Jowar-rich Multigrain Atta

Strengths

- The product is immensely liked by most of the respondents who have tasted it.
- More than 90 per cent of the survey respondents expressed their willingness to purchase the product.
- Health benefit of *jowar atta* is the major acceptance criteria in traditional areas.

- The respondents could discern no significant difference between *jowar atta* and wheat *atta*.
- *Jowar atta* is an ideal product in diet plans and weight management.
- The product is suitable for all family members, and is not only restricted to the diabetics, sick people, kids, and elderly.

Weaknesses

- Shelf-life of *jowar atta* is just one month.
- The typical *jowar* smell in the *atta* during the preparation is not liked by many consumers (especially the new consumers).
- The Delhi consumers were not happy with the way the *chapati* showed swelling atribute.
- Stickiness and texture of *jowar atta* also need to be addressed.
- The business of dealing with the *jowar atta* is at its infancy and even the supply chain is still in formative stage.
- There is a lack of professional manufacturing, marketing, and sales department
- Economy of scale, learning curve, expertise in commercial operation, and packaging are yet to be fine-tuned.

Opportunity

- *Jowar-rich* multigrain *atta* is only product of *jowar* available in the market.
- There are very few players dealing with the multigrain *atta* segment in the market.
- The multigrain *atta* consumers are not loyal to the brand as they have high expectations.
- Increasing health consciousness, disposable incomes, diminishing culinary skills, and the rising need for convenience food could be the driving factors of popularising *jowar* products in the urban traditional *jowar* consuming areas.

Threat

- Size of the multigrain *atta* industry is very small, and the market is dominated by various national-level brands.
- Growth of this segment is very less (3–4 per cent annually).
- Higher price of *jowar* seed would in turn increase the price of *jowar*-rich multigrain *atta*.

- Cost of marketing and distribution is very high for the new product.
- Use of *jowar* as a food product is restricted to the traditional consuming areas.

Jowar biscuits

Strengths

- *Jowar* biscuit is immensely liked by most of the respondents.
- More that 90 per cent of the survey respondents showed their willingness to purchase the product.
- Health benefits associated with *jowar* biscuit is the major acceptance criteria in the traditional areas.
- *Jowar* biscuit is suitable for all family members and not only restricted to diabetics, sick people, kids, and elderly.
- It is an ideal product in diet plans and weight management.
- The taste of *jowar* biscuits is not very much different from that of the existing cookies.

Weaknesses

- The shelf life of the product is very short (one month).
- Typical smell of *jowar* in the biscuit is not liked by some consumers.
- Coconut biscuits are less crispy, while normal biscuits are hard while chewing.
- The gumminess (the property that ensures that while chewing biscuits getting broken into particles and becoming ready to swallow) in *jowar* biscuits is less as compared to the normal cookies.
- Taste and after-taste are not liked by some consumers.
- This business venture is relatively new, and the supply chain is still in the formative stage.
- Lack of professional manufacturing, marketing, and sales department.
- Lack of production and operational advantages like economy of scale, learning curve, expertise in commercial operation, packaging, etc.

Opportunity

- *Jowar* biscuit is the only *jowar* product available in the market.
- Product can compete with the other cookies targeted at the health conscious consumer segment.

- The consumers of the health biscuits are not loyal to the brand as they seek more health benefits from such type of cookies.
- Increasing health consciousness, disposable incomes, diminishing culinary skills, and the rising demand for convenience food could be driving factors of popularising *jowar* products in the urban-centric traditional areas.

Threat

- Share of cookies in the total biscuit market is very small (5–7 per cent of total).
- Healthy cookies are niche products and dominated by the established players or brands.
- Growth of the cookies segment is also very less (3–4 per cent annual [CAGR 2008–13] as compared to 12–15 per cent for all) (*Source*: Datamonitor).
- Higher price of *jowar* seed would in turn increase the price of *jowar* biscuit.
- Cost of marketing and distribution is very high for the new product.
- Use of *jowar* as a food product is restricted to the traditional consuming areas.

Jowar vermicelli

Strengths

- The product is immensely liked by most of the respondents who tasted the product.
- More than 90 per cent of the survey respondents are willing to purchase the product.
- Health benefit of *jowar* is the major acceptance criterion in the traditional areas.
- Ideal product in diet plans and weight management.
- Suitable for all family members, and not restricted to only diabetics, sick people, kids, and elderly.

Weaknesses

- Shelf-life of the product is very short.
- Typical smell of *jowar* in the final form (*kheer*) is not liked by some consumers.
- The cooked item leaves a granular feeling in mouth.

- The final product is lumpy and less sticky.
- Colour of the product is darker than the wheat vermicelli.
- Roasting of the vermicelli requires more oil/fat.

Opportunity

- *Jowar*-based vermicelli is a unique product available in the market.
- The competitors are partially organised but are scattered.
- Increasing health consciousness, disposable incomes, diminishing culinary skills, and the rising demand for convenience food could be driving factors of popularising *jowar* products in urban-centric traditional areas.
- Good growth in vermicelli market (15–20 per cent per annum).

Threat

- The regional players dominate the market.
- Range of value-added products available (rice, wheat, roasted, broken, long, etc.).
- Consumer preference to the type of vermicelli (thin, roasted, long, broken, etc.) varies with regions.
- Higher price of *jowar* seed would in turn increase the price of *jowar* vermicelli.
- Cost of marketing and distribution is very high for the new product.
- Use of *jowar* as a food product is restricted to the traditional consuming areas.

Jowar suji/rawa/semolina

Strengths

- *Jowar rawa* is similar in appearance to that of wheat *rawa*.
- More than 70 per cent of the survey respondents are willing to purchase the product.
- Health benefit associated with *jowar* is the major acceptance criteria in traditional areas.
- Suitable for all family members, and not restricted to only diabetics, sick peole, kids, and elderly.
- Ideal product in diet plans and weight management.
- Taste of *jowar rawa* is not much different from the existing *suji/rawa/* semolina.

Weaknesses

- The shelf-life of the product is very short.
- Particle size of *rawa* is bolder compared to normal *rawa*.
- Water absorption capacity of *jowar rawa* is poor as it requires more water compared to wheat *rawa*.
- It is not sticky.
- Overcooking makes the final product lumpy. Roasting require more fat (oil/ghee).
- A feeling of roughness on the tongue after chewing cooked *rawa*. The granule remains hard (less water) or becomes hard soon (more water) in cooked form.
- A relatively new business venture, and the supply chain is still in the formative stage.

Opportunity

- Jowar *rawa* can be targeted at the health conscious consumer segment.
- The *rawa* segment is highly unorganised as no branded players exist in the market.
- Size of the market is comparatively large.
- Product can compete with regular *rawa* when targeting the health conscious consumer segment.
- Increasing health consciousness, disposable incomes, diminishing culinary skills, and the rising need for convenience food could be driving factors of popularising the *jowar* products in the urban-centric traditional areas.

Threat

- Highly price-sensitive market; product is manufactured locally and mainly sold in loose form.
- Growth of the *rawa* market is moderate (4–5 per cent annually) (*Source*: Foretell Internal Research).
- Higher price of *jowar* seed would in turn increase the price of *jowar rawa*.
- Cost of marketing and distribution is very high for non-popular type of product.
- Use of *jowar* as food product is restricted to the traditional consuming areas.

Porter's 5 Force Model

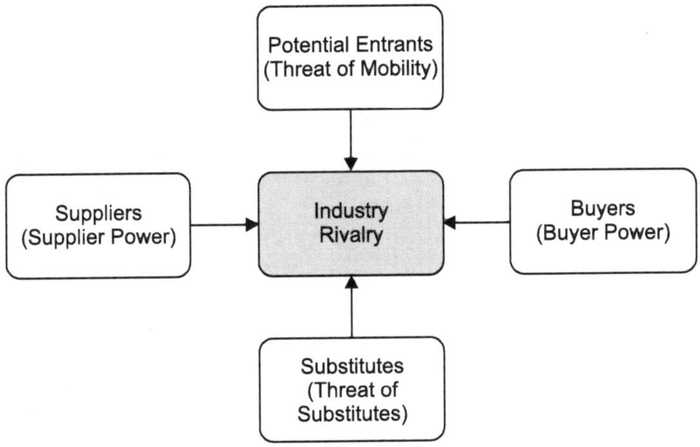

Fig. A11: Porter's 5 force model

Threat of new entrants: Medium

The segment is small but promising. There is a substantial retail margin of 20–25 per cent (10–12 per cent for regional players) for the established players (multigrain biscuits, *atta*, noodles). *Jowar suji/rawa*/semolina and *atta* are already being marketed and sold online at a very small scale (Delhi and Kolkata) as gluten-free product.

- www.health-shoppe.com,
- www.wellnessqed.com,
- ww.sunirafoods.com, http://www.energyhome.in/index.html,
- http://www.miltopexports.com/index.html

Bargaining power of supplier: Low

The brand is at its infancy and a recent entry to the market. There is a lack of commercial knowledge. There is a high demand for the new product which is sold by organised retailers,

Bargaining power of buyer: High

There is a lack of awareness about the product among consumers. The brand is being promoted by the government promoted brand among the mostly first time users in non-traditional areas and occasionally in urban-centric traditional areas.

Threat of substitute product: Medium

There are very few manufacturers selling *jowar atta* and *suji*/*rawa*/semolina. No other brand/company has come out with the variety of the products made from *jowar* because of costly and lengthy research and development required for establishing the product in the market.

Rivalry among the competitors: Low

Business operation is of very small size and players are scattered. Only multigrain *atta* is currently being marketed by organised and established players.

ANNEXURE VII

Name of the Areas Covered in Each City

Hyderabad

Maredpally	Chintalabasthi	GaddiAnnaram	Amberpet
Seethaphalmandi	Mehdipatnam	Nacharam	Ameerpet
A. G. Colony	Rajendranagar	Old Malakpet	Banjarahills
Ashoknagar	Ramanthapur	P.T.Colony	Domalguda
Chikkadapally	Redhills	Saidabad	Erragadda
Markandeyanagar	Vidyanagar	Santoshnagar	Ghandhinagar
Tallagadda	Yousufguda	Somajiguda	Jublihills
Vengalraonagar	Nallakunta	Khairtabad	Kachiguda

New Delhi

Pitampura	Malviya Nagar	Moti Nagar	Tilak Nagar	Laxmi Nagar
Rohini Sec-8	Lajpat Nagar	Patel Nagar	Subhash Nagar	Dilshad Garden
Lawrence Road	Kalkaji	Gole Market Connaught Place	Vikaspuri	Geeta Colony
Model Town-II	R.K Puram	Rajendra Nagar	Janakpuri	Yamuna Vihar
Shalimar Bagh	Sarojini Nagar	Aram Bagh	PaschimVihar	Jhilmil Colony

Pune

Aundh	Chandannagar	Bund Garden	Koregaon Park
Wadgaon	Hadapasar	Shivajinagar	K Nagar
Peth areas	Dhankwadi	Sheri	V Nagar
Bibewadi	Kothrud	Camp	Shankar Sheth Road
Yerwada	Bopodi	Sasane Nagar	

ANNEXURE VIII

Statistical Analysis Description

Factor Analysis

A common rule of thumb for dropping the least important factors from the analysis is the K1 rule. Factors with more than 1.0 as Eigen value were selected. A Cattell Scree test plot tool was also drawn to justify the extent of component to be selected.

On the benchmark measuring point of 0.600, the reliability of the data and consistency in the responses received for factors were explained as at what extent the data is reliable or consistent.

The test could not be conducted for those factors that consist of independent variables and responses in different scale or binominal.

Descriptions of Factors/Dimensions

Organoleptic: This is the most important dimension identified by the respondent. This deals with the physical attributes and organoleptic characteristics of the product. Physical attributes include product appearance in raw form as well as during the preparation (cooking) form and finished form. Organoleptic attributes also include sensory aspects (cooked), appearance, test, aroma, texture, etc.

Prior awareness on health and other benefit of *jowar*: This dimension includes awareness about the consumption of *jowar* and benefits/perceptions attached with it. Awareness on nutritional benefits, ease of digestion, good source of energy, modern food, etc., are the parameters grouped in this dimension.

Comparison of *jowar*-rich multigrain *atta* with traditional wheat *atta*: The comparison of *jowar*-rich multigrain *atta* with traditional wheat on the basis of difficulties in preparation process and taste is the major attribute included under this dimension.

Socio-economic and value for money: Economic aspects of the respondents like income, expenditure on food, and product pricing are the important aspects that are covered under this dimension.

Demography: Family size, number of children, family background, and occupations of the respondents are the attributes included in this dimension.

Myths Associated with *Jowar*: In many areas, consumers have wrong perceptions that *jowar* should be restricted only to the old, sick, and adult members of the family. This is also another dimension of the consumer prejudice on *jowar* consumption.

Cluster Analysis

Cluster analysis was carried to classify the consumer segments based on their willingness to purchase the product. Following this, the clusters were further grouped as 'Good', 'Average', and 'Bad' clusters for clarity in the output. Each cluster is compared and classified based on the analysis of the consumer's feedback through three parameters, namely, percentage of purchase vs. non-purchase out of total, attribute likewise ranking, and perceptions on comparison with the traditional/current consuming products.

The important characters of both Good and Average clusters are presented in the tables and the output of the cluster is further statistically tested (Chi-square and Probability (*P* value) to know the significant similarity and difference between both the clusters.

Similarities and Dissimilarities in the Response

Paired comparison test was conducted using ANOVA techniques. Result of this analysis is helpful in deciding communication strategy to market the *jowar*-rich multigrain *atta* in a particular city (* 95 per cent confidence level (<0.05)).

Product Improvement Variables

A full variable loaded logistic regression model was built to find out less significant variables/attributes as quantitative method. Odds Ratio Score (ODS) estimate analysis was conducted to find out low scored variables/attributes, and through analysis of Maximum Likelihood Estimates (MLE) probability of each variable was calculated. Each variable/attribute was analysed and less significant or non-significant variables were rejected (low ODS variable compared with probability to find out less significant variable with maximum level of confidence).

ANNEXURE IX

Firms Currently Marketing *Jowar*-based Products

Sunira Food Products Pvt. Ltd

2, Rowland Road, Calcutta-700020,
West Bengal
Tel.: 033-40082638; Mobile: 09836909117; Fax: 03324745634
http: //www.sunirafoods.com; E-mail: sunirafoods@gmail.com

Wellness Q. E. D.

Mr Sanjay Tiwari (CEO)
No. 322, Neb Sarai
New Delhi, Delhi-110 068, India
Tel.: +(91)-(11)-29531861; Mobile: +(91)-9810006803
E-mail: info@wellnessqed.com

Morarka Organic Foods Pvt. Ltd-Organic Puffed Jowar

Morarka Organic Foods Pvt. Ltd
Vatika Road, Off Tonk Road, Post-Shree Ram Ki Nangal,
Via-Sitapura RIICO, Jaipur-302022, Rajasthan, India.
Tel.: 91-141-2770089; Fax: 91-141-2770031
E-mail: info@morarkamail.com

Viraki Brothers

280-282, Satyam House, Narshi Natha Street, Bhat Bazar, Mumbai-400009
Mumbai, Maharashtra, India-400009
Tel.: 022-3757788; Fax: 022-4165401
http: www.seekandsource.com/viraki

Bibliography

AGMARKNET. 2002. *Marketable Surplus and Post Harvest Losses of Jowar in India.* Nagpur: Ministry of Agriculture, Directorate of Marketing and Inspection, Government of India. MRPC-45

Alfieri, M.A.H., Pomerleau, J., Grace, D.M., Anderson, L. 1995. 'Fiber Intake of Normal Weight, Moderately Obese and Severely Obese Subjects.' *Obesity Research* 3(6): 541–47

Ali, R., Staub, J., Leveille, G.A., Boyle, P.C. 1982. 'Dietary Fiber and Obesity.' In *Dietary Fiber in Health and Disease*, edited by G.V. Vahouny and D. Kritchevsky. New York: Plenum Press

Anderson, J.W. 2003. 'Whole Grains Protect Against Atherosclerotic Cardiovascular Disease.' *Proceedings of the Nutrition Society* 62(01): 135–42

Anderson, J.W. and Hanna, T.J. 1999. 'Whole Grains and Protection Against Coronary Heart Disease: What Are the Active Components and Mechanisms.' *American Journal Clinical Nutrition* 70: 307–08

Asp N.G. 1994. 'Nutritional Classification and Analysis of Food Carbohydrates.' *American Journal of Clinical Nutrition* 59(1): 679S–81S

Awika, J.M. and Rooney, L.W. 2004. 'Sorghum Photochemicals and Potential Impact on Human Health.' *Phytochemistry* 65: 1199–1221

Ayamdoo, J. A., Demuyakor, B., Dogbe, W., Owusu, R. 2013. 'Parboiling of Paddy Rice: The Science and Perceptions of it as Practiced in Northern Ghana.' *International Journal of Scientific and Technology Research* 2(4): 13–18

Ayo, J.A. and Gaffa, T. 2002. 'Effect of Undefatted Soyabean Flour on the Protein Content and Sensory Quality of "KunnuZaki" Niger.' *Food Journal* 20:7–9

Ayyangar, G.N.R. and Ayyer, M.A.S. 1936. 'Sorghum for Popping.' *Madras Agricultural Journal* 25:323–28

Basavaraj, G. and Rao, P.P. 2011. 'Regional Analysis of Household Consumption of Sorghum in Major Sorghum-Producing and Consuming States in India.' Working Paper Series no. 28. International Crops Research Institute for the Semi-Arid Tropics, Patancheru, Andhra Pradesh

Burkitt, D.P. and Trowell, H.C. 1975. *Refined Carbohydrate Foods and Disease*. London, New York, San Francisco: Academic Press

Carolina, C., Luigi, M., Nicola, C., Cristina, B., Luigi, D.G., Domenica, R.M., Paola, P., Natale, F., Scott, R.B., Brian, L., Marco, L. 2007. 'Celiac Disease: In Vitro and In Vivo Safety and Palatability of Wheat-free Sorghum Food Products.' *Clinical Nutrition* 26: 799-805

Carr, Timothy P., Curtis, L.W., Vicki, L.S., Susan, L.C., David, M.G., Kyle, R.J. 2005. 'Grain Sorghum Lipid Extract Reduces Cholesterol Absorption and Plasma Non-HDL Cholesterol Concentration in Hamsters.' *Journal of Nutrition* 135(9): 2236-40

Chand, R. 2007. 'Demand for Food Grains.' *Economic and Political Weekly* 42(52): 10–13

Channal, G., Nagnur, S., Nanjayyanamath, C. 2004. 'Indigenous Grain Storage Structures.' *Leisa India* 6(3): 10

Dayakar Rao, B., Patil, J.V., Nirmal Reddy, K., Talwar, H.S., Varma, V., Sriharsha, B., Kachui, N. 2010a. *Recent Trends in Production, Utilization and Trade of Sorghum in India*. Hyderabad: Directorate of Sorghum Research

Dayakar Rao, B., Patil, J.V., Rajendraprasad, M.P., Nirmal Reddy, K., Devi, K., Sriharsha, B., Kachui, N. 2010b. 'Impact of Innovations in Value Chain on Sorghum Farmers.' *Agricultural Economics Research Review* 23: 419-26

Dayakar Rao, B., Seetharama, N., Suresh, A., Sreekanth, M., Reddy, K.N., Rao, S.V. 2010c. *Dynamics of Value, and Trade Channels of Sorghum in India*. Rajendranagar, Hyderabad: Directorate of Sorghum Research

Desikachar, H.S.R. 1982. 'Pearling and Milling Studies on Sorghum.' In *Proceedings of the International Symposium on Sorghum Grain Quality*, edited by L.W. Rooney and D.S. Murty, Hyderabad, Inde, 28-31 October 1981, p. 194-99. Patancheru: ICRISAT

Dykes, L. and Rooney, L.W. 2006. 'Sorghum and Millet Phenols and Antioxidants.' *Cereal Science* 44: 236-51

Gazzaz, S.S., Rasco, B.A., Dong, F.M., Borham, M. 1989. 'Effect of Processing on the Thiamine, Riboflavin, and Vitamin B12 Content of Fermented Whole Grain Cereal Products.' *Journal of Food Processing and Preservation* 13: 321-34

Millet Network of India, National Institute of Rural Development (India). 2008. 'God's own Crops. National Consultation on Millets, National Institute of Rural Development, Hyderabad June 5 & 6, 2008.' *Journal of Agriculture and Food Chemistry* 49: 1620–24.

Gomez-Cordoves, C., Bartolome, B., Vieira, W., Virador, V.M. 2001. 'Effects of Wine Phenolics and Sorghum Tannins on Tyrosinase Activity and Growth of Melanoma Cells.' *Journal of Agricultural and Food Chemistry* 49: 1620–24

Gopalan, C., Rama Sastri, B.V., Balasubramanian, S.C. 1996. *Nutritive Value of Indian Foods*. Hyderabad: National Institute of Nutrition, Indian Council of Medical Research.

Grimmer, H.R., Parbhoo, V., McGrath, R.M. 1992. 'Antimutagenicity of Polyphenol-rich Fractions from *Sorghum bicolor* Grain.' *Journal of the Science of Food and Agriculture* 59(2): 25156

Hadimani, N.A. 1994. 'Pearl Millet (*Pennisetum americannum*) Grain Quality with Special Emphasis on Milling and Popping of Millet.' PhD diss. CFTRI, Mysore, India

Heidemann, C., Schulze, M.B., Franco, O.H., van Dam, R.M., Mantzoros, C.S., Hu, F.B. 2008. 'Dietary Patterns and Risk of Mortality from Cardiovascular Disease, Cancer, and All Causes in a Prospective Cohort of Women.' *Circulation* 118: 230–37

Hosmani, M.M. and Chittapur, B.M. 1997. *Sorghum Production Technology*. Dharwad: University of Agricultural Sciences

Jensen, M.K., Koh-Banerjee, P., Hu, F.B., Franz, M., Sampson, L., Grønbæk, M., Rimm, E.B. 2004. 'Intakes of Whole Grains, Bran, and Germ and the Risk of Coronary Heart Disease in Men.' *The American Journal of Clinical Nutrition* 80(6): 1492-99

Kamaliya, K.B. and Subhash, R. 2004. 'Bakery Products as Functional Foods.' *Beverage and Food World* 32(9): 33–35.

Kulkarni, S.D. 1997. 'Roasted Soyabean in Cookies Influence on Product Quality.' *Journal of Food Science Technology* 34:503–05

Kartikeyan, C., Veeraraghavantham, D., Karpagam, D., Firdouse, S.A. 2009. 'Traditional Storage Practices.' *Indian Journal of Traditional Knowledge* 8(4): 564-68

Marsland, N. and Rao, P.P. 1999. 'Marketing of Rainy and Postrainy Season Sorghum in Andhra Pradesh, Karnataka, and Maharashtra,' Working Paper Series no. 1, International Crops Research Institute for the Semi-Arid Tropics, Patancheru, Andhra Pradesh

McDonough, C.M., Rooney, L.W., Earp, C.F. 1986. 'Structural Characteristics of Eleusinecoracana (finger millet) Using Scanning Electron and Fluorescence Microscopy.' *Food Microstructure* 5: 247–56

McKeown, N.M., Meigs, J.B., Liu, S., Saltzman, E., Wilson, P.W., Jacques, P.F. 2004. 'Carbohydrate Nutrition, Insulin Resistance, and the Prevalence of the Metabolic Syndrome in the Framingham Offspring Cohort.' *Diabetes Care* 27(2): 53846

Montonen, J., Knekt, P., Järvinen, R., Aromaa, A., Reunanen, A. 2003. 'Whole-Grain and Fiber Intake and the Incidence of Type 2 Diabetes.' *The American Journal of Clinical Nutrition* 77(3): 622-29

Murty, D.S. and Kumar, K.A. 1995. 'Traditional Uses of Sorghum and Millets.' In *Sorghum and Millets: Chemistry and Technology,* edited by D.A.Y. Dendyed, 185-222. Minnesota, USA: American Association of Cereal Chemists

Parbhoo, V., Grimmer, H.R., Cameron–Clarke, A., McGrath, R. 1995. 'Induction of Cytochrome P- 450 in Rat Liver by Polyphenol Rich Diet from a Bird Resistant Sorghum Rich Diet.' *Journal of the Science of Food and Agriculture* 69: 249-52

Rao, B.S.N. 2003. 'Bioactive Phytochemicals in Indian Foods and their Potential in Health Promotion and Disease Prevention.'*Asia Pacific Journal of Clinical Nutrition* 12(1): 9–22

Rooney, L.W., Faubion, J.M., Earp, C.F. 1980. 'The Technology of Sorghum Products.' In *Cereals for Food and Beverages: Recent Progress in Cereal Chemistry,* edited by G.E. Inglett and L. Munck, 513–54. New York: Academic Press

Sabry, Z.I. and Tannous, R.I. 1961. 'Effect of Parboiling on the Thiamine, Riboflavin and Niacin Contents of Wheat.' *Cereal Chemistry* 38:536–39

Sandhya, N.S., Meera, M.S., Sila, B. 2011. 'Effect of Popping on Sorghum Starch Digestibility and Predicted Glycemic Index.'*Journal of Food Science and Technology* 50(2): 387–92

Schneeman, B.O. and Tietyen, J. 1994. 'Dietary Fiber.' In *Modern Nutrition in Health and Disease*, 8th ed., edited by M.E. Shills, J.A. Olson, and M. Shike, 89–100. Philadelphia, PA: Lea and Febiger

Shandera, D.L. Parkhurst, A.M. Jackson, D.S. 1995. 'Interactions of Sulfur Dioxide, Lactic Acid, and Temperature during Simulated Corn Wet Milling.' *Cereal Chemists* 72(4): 371–78

Shinde, R.K.U. 2004. 'Utility of Bran Products in Non Insulin Dependant Diabetes Mellitus (NIDDM) Patients.'*Journal of Human Ecology* 16(3): 219–22

Singh, P.K. 2010. 'A Decentralized and Holistic Approach for Grain Management in India.'*Current Science* 99(9): 1179–80

Slavin, J. 2003. 'Why Whole Grains are Protective: Biological Mechanisms.'*Proceedings of the Nutrition Society* 62(01): 129–34

Slavin, J. and Slavin, J. 2004. 'Whole Grains and Human Health.'*Nutrition Research Reviews* 17(1): 99–110

Subramanian, A. 1956. 'Jowar Pops are a Real Delicacy.' *Indian Farming* 6: 25–26

Subramanian, V., Butler, L.G., Jambunathan, R., Rao, K.E.P. 1983. 'Some Agronomic and Biochemical Characters of Brown Sorghums and their Possible Role in Bird Resistance.'*Journal of Agricultural Food Chemistry* 31:1303–07

Taylor, J.R.N. and Schussler, L. 1986. 'The Protein Compositions of the Different Anatomical Parts of Sorghum Grain.' *Journal of Cereal Science* 4:361–69

Ulrich, K., Ravi, S.B., Rao, B.D., Yogan, B. 2007. Industrial Utilization of Sorghum in India. *SATeJournal* 3(1)

Van Rensburg, S.J. 1981. 'Epidemiological and Dietary Evidence for a Specific Nutritional Predisposition to Esophageal Cancer.'*Journal of the National Cancer Institute* 67: 243–51

Vimala, V., Geervani, P., Uma, P., Ramadevi. 1996. 'Effect of Dehulling and Genotypes of Sorghum (*Sorghum vulgare*) on *Roti* Quality.' *Journal of Food Science and Technology* 33(3): 234–36

Warsi, A.S., Wright, B.C. 1973. 'Sorghum Growth and its Composition in Relation to Rates and Methods of Nitrogen Application. II - Pattern of N Accumulation.' *Indian Journal of Agronomy* 18(3): 277–81.

World Bank Report. 1999. 'Post-Harvest Management – Fights hunger with FAO.' *India Grains, March 2002* 4(3): 20–22.

Wursch, P. 1997. 'The Role of Viscous Soluble Fiber in the Metabolic Control of Diabetes.' *Diabetes Care* 20(11): 1774-80

Yenagi, N.B., Chittapur, B.M., Kachapur, M.D., Nadagoud, V.V. 2005. 'Suitable Process for Popping of Sorghum Cultivars for Value Addition.' *Maharashtra Journal of Agricultural Extension Education*